Lecture Notes in Mathematics

A collection of informal reports and seminars
Edited by A. Dold, Heidelberg and B. Eckmann, Zürich

76

R. G. Swan
University of Chicago, Chicago, Illinois

Algebraic K-Theory

1968

Springer-Verlag Berlin · Heidelberg · New York

Library of Congress Catalog Card Number 68 - 59063 Printed in Germany. Title No. 3682

INTRODUCTION

These notes are taken from a course on algebraic K-theory which I gave at the University of Chicago in 1967. They also include some material from an earlier course on abelian categories, elaborating certain parts of Gabriel's thesis. The results on K-theory are mostly of a very general nature. I hope to treat some of the deeper parts of the theory, in particular, the case of finite groups, in a subsequent set of notes. The point of view taken in these notes is not always consistent. I generalized a number of results on modules to abelian categories but did not hesitate to return to modules when the going got a bit rough. Most of the material here has appeared in some form in the literature, the main exception being Chapter 14 which is based on unpublished results of Milnor, Kervaire, and Steinberg.

I would like to thank J. Burroughs, G. Evans, and M. Schacher for taking the notes. In particular, I would like to thank G. Evans who collected the notes, rewrote them all in readable form, and proofread the final version. Special thanks are due to S. Mac Lane who suggested to us the idea of publishing these notes and who arranged for their typing, which was done by M. Benson, and their publication.

TABLE OF CONTENTS

PART I

CATEGORY THEORY

The purpose of this section is to provide basic information about abelian categories and Serre subcategories.

Definition. Let $\underline{A}$ be an abelian category. A subcategory $\underline{C}$ is a Serre subcategory of $\underline{A}$ if

1) $\underline{C}$ is a full subcategory of $\underline{A}$

2) If $0 \to A' \to A \to A'' \to 0$ is exact in $\underline{A}$, then $A \in \underline{C}$ if and only if A' and A'' are in $\underline{C}$.

3) $\underline{C}$ is nonempty.

Examples:

1) Let $\underline{C}$ be all the weakly effaceable functors in $(\underline{A}, \underline{Ab})$, the category of functors from $\underline{A}$ to $\underline{Ab}$. (See the example before proposition 1.10 for the definition of weakly effaceable.)

2) $\underline{A} = \underline{Ab}$, the category of all abelian groups, then examples of Serre subcategories are:

a) all torsion groups,

b) all finitely generated groups,

c) all finite groups, or

d) all p-groups.

3) Let T be an exact functor from $\underline{A}$ to $\underline{B}$, two abelian categories. Let $\underline{C}$ be the full subcategory of $\underline{A}$ with objects equal to $\{A \mid T(A) = 0\}$. Then $\underline{C}$ is a Serre subcategory of $\underline{A}$.

Serre has given the following partial converse of example 3: If $\underline{A}$ is well powered and abelian, every Serre subcategory is the full subcategory of $\underline{A}$ whose objects are $\{A \mid T(A) = 0\}$ for some exact functor T. In fact, there is even a universal exact functor. That is, there exists $T: \underline{A} \to \underline{A}/\underline{C}$ an exact functor with $\underline{A}/\underline{C}$ an abelian category such that if $S: \underline{A} \to \underline{B}$ is another exact functor between abelian categories with $S(\underline{C}) = 0$, then there exists a unique exact functor $G: \underline{A}/\underline{C} \to \underline{B}$ such that the following diagram commutes:

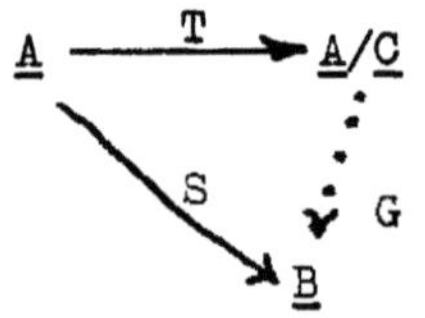

.

Proposition 1.1. A nonempty full subcategory $\underline{C}$ of an abelian category $\underline{A}$ is a Serre subcategory if and only if

$$A' \xrightarrow{f} A \xrightarrow{g} A'' \text{ an exact}$$

sequence in $\underline{A}$ with A' and A" in $\underline{C}$ implies $A \in \underline{C}$.

Proof.

$\Leftarrow$ Let $0 \to A' \to A \to A'' \to 0$ be exact in $\underline{A}$. We only need to show that $A \in \underline{C}$ implies A' and $A'' \in \underline{C}$. But $\underline{C}$ is nonempty; say $B \in \underline{C}$. Then $B \to 0 \to B$ is exact, and therefore $0 \in \underline{C}$. $0 \to A' \to A$ exact implies $A' \in \underline{C}$, and $A \to A'' \to 0$ exact implies $A'' \in \underline{C}$.

$\Rightarrow$ Let $\underline{C}$ be a Serre subcategory and let $A' \xrightarrow{f} A \xrightarrow{g} A''$ be exact with A' and $A'' \in \underline{C}$. We must show $A \in \underline{C}$.

$0 \to \ker f \to A' \to \operatorname{im} f \to 0$ and

$0 \to \operatorname{im} g \to A'' \to \operatorname{coker} \to 0$ are exact.

Hence, im f = ker g and im g $\in$ $\underline{C}$.

Finally,

$0 \to \ker g \to A \to \operatorname{im} g \to 0$ exact

implies A $\in$ $\underline{C}$. Done.

Definition. Let $\underline{C}$ be a Serre subcategory of $\underline{A}$ and let f: A $\to$ B in $\underline{A}$. Then

f: A $\to$ B is a $\underline{C}$-mono if ker f $\in$ $\underline{C}$,

f: A $\to$ B is a $\underline{C}$-epi if coker f $\in$ $\underline{C}$, and

f: A $\to$ B is a $\underline{C}$-iso if both ker f and coker f $\in$ $\underline{C}$.

Note: It will turn out that f is a $\underline{C}$ (blank) if and only if T(f) is a (blank) where T is the universal exact functor T: $\underline{A} \to \underline{A}/\underline{C}$.

Lemma 1.2. Let $A \xrightarrow{f} B \xrightarrow{g} C$ be an exact sequence of objects and maps in $\underline{A}$

a) If f and g are $\underline{C}$ $\left\{\begin{matrix}\text{mono}\\ \text{epi}\\ \text{iso}\end{matrix}\right.$, then so is gf.

b) If gf is a $\underline{C}$ mono, then f is a $\underline{C}$ mono.

c) If f is a $\underline{C}$ iso, then g is a $\underline{C}$ $\left\{\begin{matrix}\text{mono}\\ \text{epi}\\ \text{iso}\end{matrix}\right.$ if and only if gf is.

d) If g is a $\underline{C}$ iso, then f is a $\underline{C}$ $\left\{\begin{matrix}\text{mono}\\ \text{epi}\\ \text{iso}\end{matrix}\right.$ if and only if gf is.

Proof. The 3 lemma says that the following sequence is exact:

$$0 \to \ker f \to \ker gf \to \ker g \to \operatorname{coker} f \to \operatorname{coker} gf \to \operatorname{coker} g \to 0.$$

The lemma follows easily from this. Done.

Definition. Let $\underline{A}$ be an abelian category, $\underline{C}$ a Serre subcategory, and $L \in \underline{A}$. L is $\underline{C}$ closed if $u: A \to B$ is a $\underline{C}$ iso implies

$$\text{Hom}(A, L) \longleftarrow \text{Hom}(B, L) \text{ is an iso.}$$

Remark. The only object that clearly satisfies this condition is 0.

Lemma 1.3. L is $\underline{C}$ closed if and only if

1) L has no $\underline{C}$ subobject. That is, $C \in \underline{C}$ implies $\text{Hom}(C, L) = 0$, and

2) If $0 \to L \to X \to C \to 0$ is exact and $C \in \underline{C}$, then the sequence splits.

Proof. $\Longrightarrow$ 1) Let $C \in \underline{C}$. Then $C \to 0$ is a $\underline{C}$ iso. Therefore, Hom(C, L) is isomorphic to Hom(0, L) which is 0.

2) Let $0 \to L \xrightarrow{i} X \xrightarrow{j} C \to 0$ be exact with $C \in \underline{C}$. Then i is a $\underline{C}$ iso. Hence the map $\text{Hom}(C, L) \longleftarrow \text{Hom}(X, L)$ is an iso. Hence, there exists $p \in \text{Hom}(X, L)$ such that p goes to 1_i. That is $pi = 1_L$. Therefore, the sequence splits.

$\Longleftarrow$ Let $f: A \to B$ be a $\underline{C}$ iso. Then

$$0 \to \ker f \to A \xrightarrow{j} \text{im } f \to 0 \quad \text{and}$$

$$0 \to \text{im } f \xrightarrow{i} B \to \text{coker } f \to 0 \quad \text{are exact and}$$

i and j are $\underline{C}$ iso. We have the maps induced by i and j.

$$\text{Hom}(A, L) \longleftarrow \text{Hom}(\text{im } f, L) \longleftarrow \text{Hom}(B, L).$$

If both of the induced maps are isos, then the composite is also. So it is sufficient to do the cases where f is an epi and where f is a mono.

Notation. To simplify the proofs we shall use (X, Y) to mean Hom(X, Y).

<u>Case 1</u>. f is an epi.

Then $0 \to C \to A \xrightarrow{f} B \to 0$ is exact and $C \in \underline{C}$. We apply the functor $(\_\_, L)$ and get

$$(C, L) \leftarrow (A, L) \leftarrow (B, L) \leftarrow 0 \text{ is exact.}$$

But $(C, L) = 0$ by 1). Therefore, $(A, L) \leftarrow (B, L)$ is an iso.

<u>Case 2</u>. f is a mono.

Then $0 \to A \xrightarrow{f} B \to C \to 0$ is exact and $C \in \underline{C}$. Thus, $(A, L) \leftarrow (B, L) \leftarrow (C, L) \leftarrow 0$ is exact and $(C, L) = 0$ as before.

Let P be the pushout in the following diagram:

$$\begin{array}{ccc} A & \xrightarrow{g} & L \\ {\scriptstyle f}\downarrow & & \downarrow{\scriptstyle f'} \\ B & \xrightarrow[g']{} & P \end{array} .$$

This yields:

$$\begin{array}{ccccccccc} 0 & \to & A & \xrightarrow{f} & B & \to & C & \to & 0 \\ & & {\scriptstyle g}\downarrow & & {\scriptstyle g'}\downarrow & & \downarrow & & \\ 0 & \to & L & \xrightarrow{f'} & P & \to & C' & \to & 0 \end{array} ,$$

where the map $C \to C'$ is an iso; hence, $C' \in \underline{C}$. By 2) $0 \to L \underset{p}{\overset{f'}{\rightleftarrows}} P \to C' \to 0$ splits. That is, there exists $p: P \to L$ so that $pf' = 1_L$.

Then $pg'f = pf'g = g$. Hence

$$(A, L) \leftarrow (B, L) \text{ is onto and, therefore, an iso.}$$

Done.

<u>Definition</u>. $u: A \to L$ is called a <u>$\underline{C}$ envelope of A</u> if u is a $\underline{C}$ iso and L is $\underline{C}$ closed.

Definition. A Serre subcategory $\underline{C}$ of $\underline{A}$ is called a localizing subcategory if every object in $\underline{A}$ has a $\underline{C}$ envelope.

Remark. Gabriel defined this by considering the functor $T\colon \underline{A} \longrightarrow \underline{A}/\underline{C}$ of theorem 1.12. He defines $\underline{C}$ to be localizing if and only if T has a left adjoint.

Theorem 1.4. Let $\underline{C}$ be a localizing subcategory of $\underline{A}$. Let $\underline{L}$ be the full subcategory of all $\underline{C}$-closed objects. Then

(1) $\underline{L}$ is reflexive. That is, the inclusion $i\colon L \longrightarrow \underline{A}$ has a left adjoint $R\colon \underline{A} \longrightarrow \underline{L}$.

(2) $\underline{L}$ is abelian (but can fail to be an exact subcategory of $\underline{A}$).

(3) R is an exact functor from $\underline{A}$ to $\underline{L}$.

(4) If $\underline{A}$ is left (resp. right) complete, then so is $\underline{L}$.

Remark. The inclusion $i\colon \underline{L} \longrightarrow \underline{A}$ is left exact since it has a left adjoint R. However i and even $iR\colon \underline{A} \longrightarrow \underline{A}$ may not be exact.

Proof.

For each $A \in \underline{A}$ choose a $\underline{C}$-envelope, $u_A\colon A \longrightarrow RA$. This defines R on objects. If $L \in \underline{L}$, then $(u_A, 1)\colon (RA, L) \xrightarrow{\approx} (A, L)$. Therefore, RA represents the functor $(A, -)$ restricted to $\underline{L}$ so $\underline{L}$ is reflexive. If $f\colon A \longrightarrow A'$ in $\underline{A}$, then Rf is the unique map making the diagram

$$\begin{array}{ccc} A & \xrightarrow{f} & A' \\ {\scriptstyle u_A}\downarrow & & \downarrow \\ RA & \xrightarrow{Rf} & RA' \end{array}$$

commute.

Lemma 1.5. $iR\colon \underline{A} \longrightarrow \underline{A}$ is left exact.

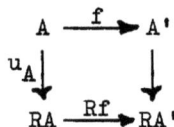

Proof.

Let $0 \to A' \xrightarrow{i} A \xrightarrow{j} A'' \to 0$ be exact in $\underline{A}$. We must show that $0 \to RA' \xrightarrow{Ri} RA \xrightarrow{Rj} RA''$ is exact in $\underline{A}$.

$RjRi = 0$ is clear since R is an additive functor.

We have the following diagram in $\underline{A}$:

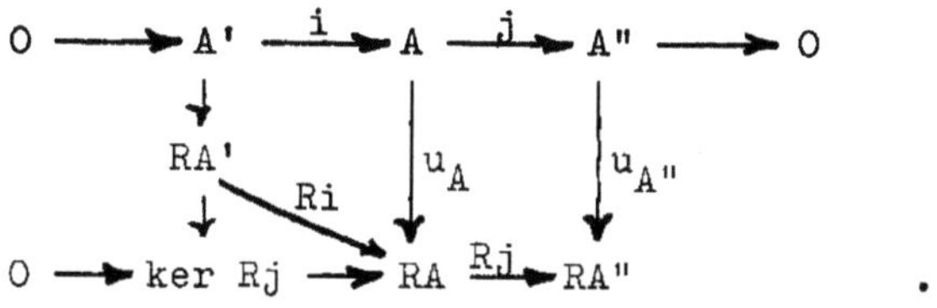

Let g be the map from A' to $\ker R_j$.

Applying the 12 lemma we obtain the exact sequence

$$0 \to \ker g \to \ker u_A \to \ker u_{A''} \to \operatorname{coker} g \to \operatorname{coker} u_A \to \operatorname{coker} u_{A''}$$

$\ker u_A$, $\ker u_{A''}$, $\operatorname{coker} u_A$, and $\operatorname{coker} u_{A''} \in \underline{C}$ imply that $\ker g$ and $\operatorname{coker} g \in \underline{C}$.

Hence $A' \to RA' \to \ker Rj$ is a $\underline{C}$ iso. Therefore,

$RA' \to \ker R_j$ is a $\underline{C}$ iso by lemma 1.2.

Thus we have an exact sequence

$0 \to C \to RA' \to \ker R_j \to D \to 0$ with C and $D \in \underline{C}$.

$C = 0$ since RA' is $\underline{C}$ closed.

But RA' $\underline{C}$ closed implies that the exact sequence $0 \to RA' \to \ker R_j \to D \to 0$ splits.

Thus, D is isomorphic to a subobject of $\ker R_j \subset RA$. But RA is $\underline{C}$ closed. Therefore, $D = 0$ and RA' is isomorphic to $\ker R_j$.

Done with lemma 1.5.

The next claim is that $\underline{L}$ is closed under products and kernels, and, hence, under $\varprojlim$ when they exist.

Let $L_\alpha \in \underline{L}$ and $L = \prod L_\alpha$ be defined in $\underline{A}$. We want that $L \in \underline{L}$. Let $u\colon A \to B$ be a $\underline{C}$ iso. Then we have

$$\begin{array}{ccc}
(A, \prod L_\alpha) & \longleftarrow & (B, \prod L_\alpha) \\
\parallel & & \parallel \\
\prod (A, L_\alpha) & \longleftarrow & \prod (B, L_\alpha)
\end{array}$$

which is an iso at each α and, hence, is an iso. Therefore, $\prod L_\alpha = L \in \underline{L}$.

Let $0 \to X \to L \to L'$ be exact in $\underline{A}$ with L and $L' \in \underline{L}$. We want $X \in \underline{L}$.

Let $u\colon A \to B$ be a $\underline{C}$ iso. Then

$$\begin{array}{ccccccc}
0 \to & (A, X) & \to & (A, L) & \to & (A, L) & \quad \text{is exact;} \\
& \uparrow & & \uparrow & & \uparrow & \\
0 \to & (B, X) & \to & (B, L) & \to & (B, L') & \quad \text{is exact;}
\end{array}$$

and $(B, L) \to (A, L)$ and $(B, L') \to (A, L')$ are isos. The five lemma proves that $(B, X) \to (A, X)$ is an iso. Hence $X \in \underline{L}$.

Now for sums and cokers. Finite sums of objects of $\underline{L}$ are in $\underline{L}$ since they are products.

Let $F\colon J \to \underline{A}$ with $\varinjlim F = A$ existing in $\underline{A}$. Then $R(\varinjlim F) = \varinjlim RF$ in $\underline{L}$. Finite $\varinjlim$ exist in $\underline{A}$. Pick $F\colon \underline{I} \to \underline{L}$ where $\underline{I}$ has only finitely many objects. Let $i\colon \underline{L} \to \underline{A}$ be the inclusion functor. $\varinjlim iF$ exists in $\underline{A}$ so $\varinjlim RiF$ exists in $\underline{L}$. Since Ri is the identity, $\varinjlim F$ exists in $\underline{L}$.

This also shows that if A is right complete, then so is $\underline{L}$.

Now $\underline{L}$ is additive and has finite $\varinjlim$ and $\varprojlim$. Therefore, finite direct sums, kernels, and cokernels exist in $\underline{L}$. To show $\underline{L}$ is abelian we need only show that coim f $\xrightarrow{\approx}$ im f for all maps f in $\underline{L}$. To do this and to show R is exact, it will suffice to show that R preserves kernels and cokernels. It will then follow that R preserves images and coimages. If f: A $\longrightarrow$ A' in $\underline{L}$ we can thus compute its image and coimage by applying i, computing im f, coim f in $\underline{A}$ and applying R. Since coim f $\xrightarrow{\approx}$ im f in $\underline{A}$ the same will hold in $\underline{L}$.

Now R clearly preserves cokernels, being a left adjoint. Also i preserves kernels, being a right adjoint. If f: A $\longrightarrow$ A' in $\underline{A}$, let K = ker f, $0 \longrightarrow K \longrightarrow A \longrightarrow A'$. By lemma 1.5, we have

$$0 \longrightarrow iRK \longrightarrow iRA \longrightarrow iRA'.$$

But i is just the inclusion and preserves kernels so $0 \longrightarrow RK \longrightarrow RA \longrightarrow RA'$ is exact in $\underline{L}$, so RK is the kernel of Rf. This shows that R preserves kernels and concludes the proof of the theorem.

We now investigate conditions under which a Serre subcategory is localizing. The methods used will also enable us to identify $\underline{L}$ in some cases.

<u>Notation</u>. If $\underline{A}$ is abelian and $\underline{C}$ is a Serre subcategory, we denote by $\underline{M}$ the full subcategory of all M $\in$ $\underline{A}$ such that M has no subobject other than 0 belonging to $\underline{C}$.

<u>Lemma 1.6</u>. If $0 \longrightarrow A \xrightarrow{f} L \xrightarrow{g} B \longrightarrow 0$ is exact in $\underline{A}$ and L $\in$ $\underline{L}$,

then $A \in \underline{L}$ if and only if $B \in \underline{M}$.

Proof. $\Longleftarrow$ Assume $B \in \underline{M}$. $A \in \underline{M}$ since $L \in \underline{L}$. By lemma 1.3 it remains to show that if $0 \longrightarrow A \xrightarrow{i} X \xrightarrow{j} C \longrightarrow 0$ is exact with $C \in \underline{C}$, then the sequence splits.

i is a $\underline{C}$-iso. Therefore $(X, L) \xrightarrow{\cong} (A, L)$. Hence $f = gi$ for a unique $g \in (X, L)$. Consider the commutative diagram in $\underline{A}$:

$$\begin{array}{ccccccccc}
0 & \longrightarrow & A & \xrightarrow{i} & X & \xrightarrow{j} & C & \longrightarrow & 0 \\
 & & {\scriptstyle 1_A}\downarrow & & {\scriptstyle g}\downarrow & & {\scriptstyle h}\downarrow & & \\
0 & \longrightarrow & A & \xrightarrow{f} & L & \xrightarrow{\chi} & B & \longrightarrow & 0
\end{array}$$

$h: C \longrightarrow B$ is the 0 map since $C \in \underline{C}$ and $B \in \underline{M}$. Therefore $\chi g = 0$ and $g: X \longrightarrow L$ factors through the kernel of χ via $X \xrightarrow{k} A \xrightarrow{f} L$.

k splits i. Because $g = fk$ and

$$f = gi = fki.$$

But f is a mono. Therefore, $ki = 1_A$.

$\Longrightarrow$ Assume $A \in \underline{L}$. Let C be a $\underline{C}$ subobject of B. Then there exists a mono $i: C \longrightarrow B$. Consider the diagram where P is the pullback:

$$\begin{array}{ccccccccc}
0 & \longrightarrow & A & \longrightarrow & L & \longrightarrow & B & \longrightarrow & 0 \\
 & & {\scriptstyle 1_A}\uparrow & & \uparrow{\scriptstyle j} & & \uparrow{\scriptstyle i} & & \\
0 & \longrightarrow & A & \longrightarrow & P & \longrightarrow & C & \longrightarrow & 0
\end{array}$$

j is a mono since i is. $A \in \underline{L}$ implies that $A \longrightarrow P \longrightarrow C$ splits. Therefore C is isomorphic to a subobject of P. But L is $\underline{C}$ closed implies that $C = 0$. Thus, $B \in \underline{M}$. Done.

Proposition 1.7. Let $\underline{C}$ be a Serre subcategory of an abelian category $\underline{A}$. Then the following are equivalent:

1) $\underline{C}$ is localizing.

2) Every $A \in \underline{A}$ has a $\underline{C}$ mono $u: A \to L$ with L $\underline{C}$ closed.

3) Every $A \in \underline{A}$ has a largest $\underline{C}$ subobject; and if $A \in \underline{M}$, then there exists a mono $A \to L$ with L $\underline{C}$ closed.

Proof.

1) $\Longrightarrow$ 2) is trivial

2) $\Longrightarrow$ 3) We are given $0 \to C \xrightarrow{i} A \xrightarrow{u} L$ with L $\underline{C}$ closed and $C \in \underline{C}$. C is the largest $\underline{C}$ subobject of A. In fact, if $D \in \underline{C}$ is a $\underline{C}$ subobject of A, then there exists a mono $j: D \to A$. $uj = 0$ since L is $\underline{C}$ closed. Therefore, there exists $f: D \to C$ such that $if = j$. Hence $D \subset C$. That is, C is the largest $\underline{C}$ subobject.

Now, let $A \in \underline{M}$. Then there exists an exact sequence $0 \to C \to A \to L$ with $C \in \underline{C}$ and L $\underline{C}$ closed. But 0 is the only $\underline{C}$ subobject of A. Therefore, $C = 0$ and we have produced a mono from A to a $\underline{C}$ closed object.

3) $\Longrightarrow$ 2) Pick $A \in \underline{A}$. Let C be its largest $\underline{C}$ subobject. Then

$$0 \to C \to A \to B \to 0 \text{ is exact.}$$

We first show that $B \in \underline{M}$. Let D be a $\underline{C}$ subobject of B. Construct the pullback diagram

$$\begin{array}{ccccccccc} 0 & \to & C & \to & P & \xrightarrow{j'} & D & \to & 0\\ & & \downarrow{\scriptstyle 1_C} & \overset{f}{\swarrow} & \downarrow{\scriptstyle \gamma'} & & \downarrow{\scriptstyle \gamma} & & \\ 0 & \to & C & \xrightarrow{i} & A & \xrightarrow{j} & B & \to & 0 \end{array} .$$

j' is an epi since j is. C and D $\in \underline{C}$ implies P $\in \underline{C}$. But C is the largest C subobject of A. Therefore, there exists f: P $\longrightarrow$ C so that $\chi' = if$. Therefore, $0 = j\chi' = \chi j'$. Finally, χ is a mono implies $j' = 0$. Thus, D = 0 and B $\in \underline{M}$.

B $\in \underline{M}$ implies that we can construct a mono B $\longrightarrow$ L with L a $\underline{C}$ closed object. Therefore, $0 \longrightarrow C \longrightarrow A \longrightarrow B \longrightarrow L$ is a $\underline{C}$ mono of A into a $\underline{C}$ closed object.

3) $\Longrightarrow$ 1) We have to show that given A $\in \underline{A}$ we can find a $\underline{C}$ envelope A $\longrightarrow$ L. We can assume A $\in \underline{M}$ since given A there exists a largest $\underline{C}$ subobject C and

$$0 \longrightarrow C \longrightarrow A \longrightarrow B \longrightarrow 0 \text{ is exact with } B \in \underline{M}.$$

If B $\longrightarrow$ L is a $\underline{C}$ iso with L $\underline{C}$-closed then A $\longrightarrow$ B $\longrightarrow$ L is a $\underline{C}$ iso with L $\underline{C}$ closed. Assume A $\in \underline{M}$.

There does exist a mono A $\longrightarrow$ L where L is $\underline{C}$ closed. This can fail to be a $\underline{C}$ iso. Let $0 \longrightarrow A \longrightarrow L \longrightarrow Q \longrightarrow 0$ be exact. Then Q has a largest $\underline{C}$ subobject, C. Form the pullback diagram

$$\begin{array}{ccccccccc} 0 & \longrightarrow & A & \longrightarrow & L & \longrightarrow & Q & \longrightarrow & 0 \\ & & {\scriptstyle 1_A}\uparrow & & \uparrow{\scriptstyle f} & & \uparrow{\scriptstyle i} & & \\ 0 & \longrightarrow & A & \longrightarrow & P & \longrightarrow & C & \longrightarrow & 0 \end{array} .$$

We claim P is $\underline{C}$ closed, and, hence, A $\longrightarrow$ P is a $\underline{C}$ iso with a $\underline{C}$ closed object.

Consider the diagram

$$\begin{array}{ccccccccc} 0 & \longrightarrow & P & \xrightarrow{f} & L & \longrightarrow & \operatorname{coker} f & \longrightarrow & 0 \\ & & \downarrow & & \downarrow & & \downarrow & & \\ 0 & \longrightarrow & C & \xrightarrow{i} & Q & \longrightarrow & \operatorname{coker} i & \longrightarrow & 0 \end{array} .$$

coker f $\longrightarrow$ coker i is an iso by the 12 lemma. Coker i $\in$ $\underline{M}$ since C is the largest $\underline{C}$ subobject of Q. Therefore by lemma 1.6 P $\in$ $\underline{L}$. That is P is $\underline{C}$ closed and $\underline{C}$ is localizing. Done.

Corollary 1.8. If $\underline{A}$ has injective envelopes, then $\underline{C}$ is localizing if and only if every A $\in$ $\underline{A}$ has a largest $\underline{C}$ subobject.

Proof. All that remains to be proved is that if A $\in$ $\underline{M}$ then there exists a mono u: A $\longrightarrow$ L with L $\underline{C}$ closed.

Let L be the injective envelope of A. We claim L $\in$ $\underline{M}$. Let C $\in$ $\underline{C}$ be a nonzero $\underline{C}$ subobject of L. Then C meets A nontrivially. But this contradicts the fact that A $\in$ $\underline{M}$.

Remark. This shows that $\underline{M}$ is closed under taking essential extensions.

Finally we need that if $0 \longrightarrow L \longrightarrow X \longrightarrow C \longrightarrow 0$ is exact and C $\in$ $\underline{C}$, then the sequence splits. This is trivial since L is injective. By lemma 1.3, L is $\underline{C}$ closed, and by proposition 1.7 $\underline{C}$ is localizing. Done.

Corollary 1.9. If $\underline{A}$ has injective envelopes, is right complete, and well powered, then $\underline{C}$ is localizing if and only if $\underline{C}$ is closed under direct sums (if and only if $\underline{C}$ is closed under $\varinjlim$).

Proof. $\Longleftarrow$ Let A $\in$ $\underline{A}$. We will show A has a largest $\underline{C}$ subobject. A has a set of $\underline{C}$ subobjects, $C_\alpha \xrightarrow{1_\alpha} A$ since $\underline{A}$ is well powered. $\sum C_\alpha \in \underline{C}$ since $\underline{C}$ is closed under $\sum$. But the image of

$$\sum C_\alpha \xrightarrow{\sum i_\alpha} A$$ is in $\underline{C}$ since $\underline{C}$ is a Serre subcategory,

and the image is clearly the largest $\underline{C}$ subobject of $\underline{A}$.

$\Longrightarrow A \in \underline{C}$ if and only if $A \longrightarrow 0$ is a $\underline{C}$ envelope if and only if $RA = 0$ where R is the reflection $R\colon \underline{A} \longrightarrow \underline{L}$.

Let $C_\alpha \in \underline{C}$. Then $\sum C_\alpha \in \underline{C}$ if and only if $R(\sum C_\alpha) = 0$. But $R(\sum C_\alpha) = \sum RC_\alpha = 0$ since R is a left adjoint. Therefore, $\underline{C}$ is closed under direct sums. Done.

Example. Let $\underline{A}$ be a small abelian category, $(\underline{A}, \underline{Ab})$ the category of covariant functors $F\colon \underline{A} \longrightarrow \underline{Ab}$. We call a functor $F \in (\underline{A}, \underline{Ab})$ weakly effaceable if for all $A \in \underline{A}$ and each $x \in F(A)$, there is a mono $0 \longrightarrow A \xrightarrow{f} B$ such that $F(f)x = 0$. Let $\underline{C}$ be the full subcategory of weakly effaceable functors. Then $\underline{C}$ is localizing. It is trivial to verify that $\underline{C}$ is a Serre subcategory. It is well known that $(\underline{A}, \underline{Ab})$ is well powered, right complete, and has injective envelopes.

We want to show that if $(W_s)_{s \in S}$ are weakly effaceable, then so is $\sum_{s \in S} W_s$. That is, given $x \in \sum_{s \in S} W_s(A)$ we want to find a mono, $0 \longrightarrow A \xrightarrow{f} B$, such that x goes to 0 under f. There exists a finite set $S' \subset S$ and a $y \in \sum_{s \in S'} W_s(A)$ such that under the map

$$\sum_{s \in S'} W_s(A) \xrightarrow{i_A} \sum_{s \in S} W_s(A)$$

y goes to x.

$\sum_{s \in S'} W_s \in \underline{C}$ since $\underline{C}$ is a Serre subcategory. Thus there exists a mono $0 \to A \xrightarrow{f} B$ effacing y. The following diagram commutes:

$$\begin{array}{ccc} \sum_{s \in S'} W_s(A) & \xrightarrow{i_A} & \sum_{s \in S} W_s(A) \\ \downarrow{\scriptstyle f_{S'}} & & \downarrow{\scriptstyle f_S} \\ \sum_{s \in S'} W_s(B) & \xrightarrow{i_B} & \sum_{s \in S} W_s(B) \end{array} .$$

y is effaced by $f_{S'}$. Therefore, $i_B f_{S'}$ is 0 on y. Therefore, $f_S i_A$ is 0 on y. But y goes to x under i_A. Thus, f_S sends x to 0, and the mono $A \xrightarrow{f} B$ effaces x.

Proposition 1.10. In the above example $\underline{L}$, the full subcategory of $\underline{C}$ closed objects, has all the left exact functors as objects.

Proof. Let $L \in \underline{L}$, let I be its injective envelope, and let $0 \to L \to I \to Q \to 0$ be exact. $I \in \underline{L}$ as in corollary 1.8. By lemma 1.6, $L \in \underline{L}$ if and only if $Q \in \underline{M}$. We will show $Q \in \underline{M}$ if and only if L is left exact. I is right exact since it is injective. In fact, if $A \to B \to C \to 0$ is exact, then $0 \to (C, -) \to (B, -) \to (A, -)$ is exact. Applying (, I) and using Yoneda's lemma gives $I(A) \to I(B) \to I(C) \to 0$ is exact. Now $I \in \underline{M}$ so I has no $\underline{C}$ subobject. Let $J(A) \subset I(A)$ consist of all effaceable elements of I(A), i.e., J(A) is the intersection of all $\ker[I(A) \to I(B)]$ for all monos $A \to B$ in $\underline{A}$. It is clear that J is weakly effaceable. Thus J = 0 so I preserves monos. Thus I is exact.

Let $0 \to A' \to A \to A'' \to 0$ be an exact sequence in $\underline{A}$. This yields the commutative diagram

$$\begin{array}{ccccccccc}
 & & 0 & & 0 & & 0 & & \\
 & & \downarrow & & \downarrow & & \downarrow & & \\
0 & \longrightarrow & L(A') & \longrightarrow & I(A') & \longrightarrow & Q(A') & \longrightarrow & 0 \\
 & & \downarrow & & \downarrow & & \downarrow & & \\
0 & \longrightarrow & L(A) & \longrightarrow & I(A) & \longrightarrow & Q(A) & \longrightarrow & 0 \\
 & & \downarrow & & \downarrow & & \downarrow & & \\
0 & \longrightarrow & L(A'') & \longrightarrow & I(A'') & \longrightarrow & Q(A'') & \longrightarrow & 0 \\
 & & \downarrow & & \downarrow & & \downarrow & & \\
 & & 0 & & 0 & & 0 & & .
\end{array}$$

The rows are exact, and the I column is exact. We want that the L column is exact if and only if the Q column is. Apply the embedding theorem and finish with a diagram chase. Done.

Corollary 1.11. In the above situation, $\underline{L}$, the full subcategory of left exact functors, forms an abelian category.

Theorem 1.12. (Mitchell) Let $\underline{A}$ be a small abelian category. Then there exists a ring R and an exact fully faithful covariant functor T: $\underline{A} \longrightarrow \underline{M}_R$ where $\underline{M}_R$ is the category of right R modules and R homomorphisms.

Proof. There exists a contravariant functor H: $\underline{A} \longrightarrow \underline{L}$ where $\underline{L}$ is the category of left exact functors from $\underline{A}$ to $\underline{Ab}$. H is given by $H(A) = h_A$ where $h_A(B) = (A, B)$ for each $B \in \underline{A}$. h_A is left exact. H is fully faithful since (A, B) is isomorphic to (h_B, h_A) by Yoneda's lemma.

By the above corollary $\underline{L}$ is an abelian category. Thus, it makes sense to ask if H is exact. Let $0 \longrightarrow A' \longrightarrow A \longrightarrow A'' \longrightarrow 0$ be an exact sequence in $\underline{A}$. Then for all $B \in \underline{A}$

$$(A', B) \longleftarrow (A, B) \longleftarrow (A'', B) \longleftarrow 0 \text{ is exact.}$$

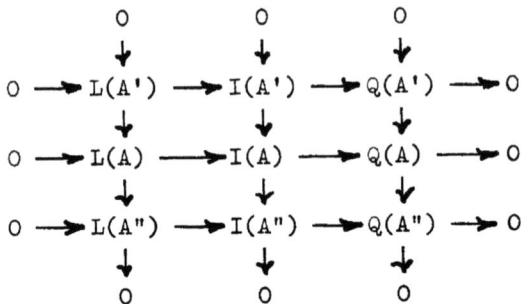

From this we produce an exact sequence in (A, $\underline{Ab}$)

$$0 \leftarrow Q \leftarrow h_{A'} \leftarrow h_A \leftarrow h_{A''} \leftarrow 0 \; .$$

We want this sequence in $\underline{L}$ so we apply the reflection functor R. h_A are all left exact so R is the identity on them. We get the following exact sequence in $\underline{L}$:

$$0 \leftarrow RQ \leftarrow h_{A'} \leftarrow h_A \leftarrow h_{A''} \leftarrow 0 \; .$$

We want that RQ = 0. That is $Q \in \underline{C}$. We must show Q is weakly effaceable.

Pick $x \in Q(B)$. We need to find a mono $B \xrightarrow{f} C$ such that $Q(f)x = 0$. Find $g \in (A', B)$ which goes to x in the map $Q(B) \leftarrow (A', B)$. Form the pushout diagram

$$\begin{array}{ccccccccc} 0 & \rightarrow & A' & \rightarrow & A & \rightarrow & A'' & \rightarrow & 0 \\ & & g\downarrow & & h\downarrow & & \downarrow 1_{A''} & & \\ 0 & \rightarrow & B & \xrightarrow{f} & P & \rightarrow & A'' & \rightarrow & 0 \end{array} \; .$$

f effaces x. Consider the diagram

$$\begin{array}{ccccccc} 0 & \leftarrow & Q(B) & \leftarrow & (A', B) & \leftarrow & (A, B) \\ & & \downarrow & & \downarrow & & \downarrow \\ 0 & \leftarrow & Q(P) & \leftarrow & (A', P) & \leftarrow & (A, P) \end{array} \; .$$

g goes to $x \in Q(B)$. g goes to fg in (A', P). But fg comes from $h \in (A, P)$, and the rows are exact. Thus, fg goes to zero in Q(P), and x goes to zero in Q(P) since the diagram commutes. Thus, x is effaced and Q is weakly effaceable.

Next we show that $\underline{L}$ has enough injectives and a generator.

Pick $A \in \underline{L} \subset (\underline{A}, \underline{Ab})$. Let $0 \rightarrow A \rightarrow I$ be the injective envelope of A in $(\underline{A}, \underline{Ab})$. $I \in \underline{L}$ as in Corollary 1.8. I is injective in $\underline{L}$ since kernels in $\underline{L}$ and in $(\underline{A}, \underline{Ab})$ are the same; and, hence, monos are the same. Thus if we have the diagram

$$\begin{array}{ccccc} & & & & I \\ & & & \nearrow & \\ 0 & \rightarrow & A & \rightarrow & B \end{array}$$

in $\underline{L}$, we have only to fill it in in $\underline{A}$. Therefore, $\underline{L}$ has enough injectives.

Let G be a generator of $(\underline{A}, \underline{Ab})$. Then RG is a generator of $\underline{L}$, i.e., (RG,) is a faithful functor on $\underline{L}$ since (RG,) = (G,) in $\underline{L}$ and (G,) is faithful in $\underline{A}$.

Setting $\underline{B}$ equal to the dual of $\underline{L}$ and F equal to the dual of H we have proved the following:

Lemma 1.13. Let $\underline{A}$ be a small abelian category. Then there exists a covariant, exact, fully faithful functor $H: \underline{A} \rightarrow \underline{B}$ where B is abelian, complete, has enough projectives, and has a cogenerator.

We shall prove theorem 1.12 using the above lemma without using what $\underline{B}$ is. First we show that under the conditions above we have the following:

Lemma 1.14. $\underline{B}$ has a projective generator.

Proof. $\underline{B}$ is well powered since it has a cogenerator. Let C be a cogenerator and let $\{A_\alpha\}$ be the set of subobjects of C. $\sum A_\alpha$ exists since $\underline{B}$ is complete. There exists a projective,

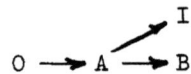

P, and an epi, $P \xrightarrow{f} \sum A_\alpha$. We claim that P is a generator. That is (P,) is a faithful functor. (P,) is exact since P is projective.

Sublemma 1.15. Let $T: \underline{C} \longrightarrow \underline{D}$ be an exact covariant functor between abelian categories $\underline{C}$ and $\underline{D}$. Then T is faithful if and only if for all $C \in \underline{C}$ $T(C) = 0$ implies $C = 0$.

Proof. $\Longrightarrow$ $T(C) = 0$ implies $T(1_C) = T(0)$. Then $1_C = 0$ implies $C = 0$ since T is faithful.

$\Longleftarrow$ We want to show $T(f) = T(g)$ implies $f = g$. We reduce to the case $T(h) = 0$ implies $h = 0$ by setting $h = f - g$ and using the fact that T exact implies T additive.

$f = 0$ if and only if im $f = 0$.

$T(\text{im } f) = \text{im } T(f)$ since T is exact.

$T(f) = 0$ implies im $T(f) = 0$. But $T(\text{im } f) = 0$ implies im $f = 0$ since T is exact. Done with lemma 1.15.

We will be done with the proof of lemma 1.14 if we show $B \in \underline{B}$ $B \neq 0$ implies $(P, B) \neq 0$. Pick $B \neq 0$. Then there exists $g: B \longrightarrow C$ such that $g = g1_B \neq 0$ since C is a cogenerator. Hence we have

$$B \xrightarrow{\text{epi}} \text{im } g \xrightarrow{\text{mono}} C.$$ Hence im g is a subobject of C. Then there exists $P \xrightarrow{\text{epi}} \text{im } g$ by projecting onto the appropriate coordinate. But P is projective so there exists $h: P \longrightarrow B$ making the following diagram commute

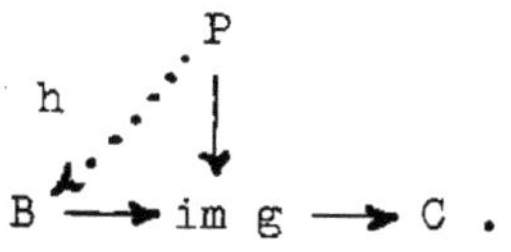

$h = 0$ implies im $g = 0$ implies $g = 0$. But $g \neq 0$ so $h \neq 0$ and $h \in (P, B)$. Hence $(P, \ )$ is faithful. Done.

Take $B \in \underline{B}$ and construct

$$\sum_{f \in (P,B)} P_f \xrightarrow{F} B \qquad \text{by taking one copy of } P,$$

the projective generator, for each $f \in (P, B)$ and map it to B by using f on P_f. We claim that F is an epi. Look at

$$* \qquad \sum_{f \in (P,B)} P_f \longrightarrow B \longrightarrow Q \longrightarrow 0 .$$

If $Q \neq 0$ then $\exists\ P \xrightarrow{h} Q$ a nonzero map. But P is projective so there exists f

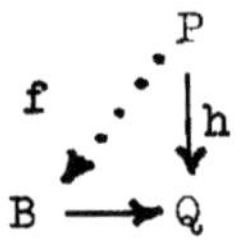

so that the diagram commutes. But the f-th component of $\sum P_f$ says that h is 0 since the sequence * is exact.

Let P_1 be any projective generator. Let

$$P = \sum_{A \in \underline{A}} \ \sum_{f \in (P_1, H(A))} P_1 .$$

Then P is a projective generator such that for each $A \in \underline{A}$ there exists an epi

$$P \longrightarrow H(A).$$

Let $R = (P, P)$ and define $\underline{B} \xrightarrow{T} \underline{M}_R$ where $\underline{M}_R$ is the category of right R modules by $T(B) = (P, B)$. We define R action by

$f: P \longrightarrow B \quad r \in R$ then

$$P \xrightarrow{r} P \xrightarrow{f} B \quad \text{(composite } fr: P \to B\text{)} .$$

That is, composition gives R action on T(B).

We have the composition of functors

$$\underline{A} \xrightarrow{H} \underline{B} \xrightarrow{T} \underline{M}_R .$$

Claim. TH is an exact, fully faithful functor.

Proof. T is exact since P is projective. TH is faithful since H is exact, fully faithful. So it only remains to check that TH is fully faithful. That is, given C and D $\in \underline{A}$ (HC, HD) $\longrightarrow$ (THC, THD) is onto. Let A = HC and B = HD. There exist epis $\mu: P \longrightarrow A$ and $\gamma: P \longrightarrow B$.

Claim. (A, B) is isomorphic to (TA, TB).

Proof. We have already checked that the induced map is 1 - 1.

Pick $f \in$ (TA, TB). That is, we have $(P, A) \xrightarrow{f} (P, B)$ an R homomorphism.

We want to find $g: A \longrightarrow B$ such that $f = (1_P, g)$. We have $\mu \in (P, A)$. μ generates (P, A) as an R module. In fact, if $x \in (P, A)$, then there exists $r: P \longrightarrow P$ so that the following diagram commutes

$$\begin{array}{ccccc} & & P & & \\ & {\scriptstyle r}\swarrow & \downarrow{\scriptstyle x} & & \\ P & \xrightarrow{\mu} & A & \longrightarrow & 0 . \end{array}$$

That is, $x = \mu r$ or μ generates (P, A) as an R module. So we need to find $g: A \longrightarrow B$ so that $f(\mu) = (1_P, g)\mu = g\mu$.

Consider the diagram

$$\begin{array}{ccccccccc} & & P & & & & & & \\ & & \downarrow h & \searrow r & & & & & \\ 0 & \longrightarrow & K & \xrightarrow{i} & P & \xrightarrow{\mu} & A & \longrightarrow & 0 \\ & & & & \downarrow 1_P & & \downarrow g & & \\ & & & & P & \xrightarrow[f(\mu)]{} & B & & \end{array}$$

If $f(\mu)1_P(K) = 0$, then g exists since K is the cokernel of i. Suppose $f(\mu)1_P i \neq 0$. Then there exists $h: P \longrightarrow K$ such that $f(\mu)1_P ih \neq 0$. But $ih: P \longrightarrow P$ and hence is an element r of R. Therefore we have $f(\mu)r \neq 0$ but $\mu r = \mu ih = 0$. f is an R homomorphism. Therefore $f(\mu)r = f(\mu r) = 0$ which is the desired contradiction. Thus g exists and the map $(A, B) \longrightarrow (TA, TB)$ is an iso. Done.

The following result shows the relation between Gabriel's and Freyd's treatment of localization.

<u>Proposition 1.16</u>. Let $\underline{A}$ be an abelian category which is well powered, left complete, and has injective envelopes. Let $\underline{M}$ be a full subcategory which is closed under subobjects, products (including infinite ones), and essential extensions. Define $\underline{C} = \{C \mid (C, M) = 0 \text{ for all } M \in \underline{M}\}$. Then $\underline{C}$ is a localizing subcategory. Conversely, given a localizing subcategory $\underline{C}$ let $\underline{M} = \{M \mid (C, M) = 0 \text{ for all } C \in \underline{C}\}$. Then $\underline{M}$ is closed under subobjects, products, and essential extensions. Furthermore, these correspondences are 1 - 1.

<u>Proof</u>. Let $\underline{C}$ be a localizing subcategory. Construct $\underline{M}$.

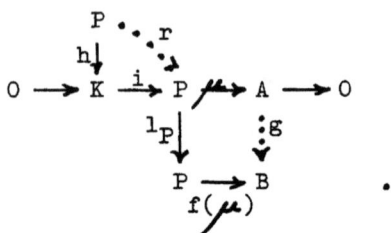

Say $M \in \underline{M}$. Let $0 \longrightarrow A \xrightarrow{i} M$ be a subobject of M. Let $f: C \longrightarrow A$ be a nonzero map with $C \in \underline{C}$. Then $if: C \longrightarrow M$ is a nonzero map which is impossible. Hence $A \in \underline{M}$ and $\underline{M}$ is closed under subobjects.

Let $M_\alpha \in M$ then $(C, \prod M_\alpha) = \prod (C, M_\alpha) = 0$. Hence, $\underline{M}$ is closed under products.

Let $0 \longrightarrow M \xrightarrow{i} A$ be an essential extension of M where $M \in \underline{M}$. Let $f: C \longrightarrow A$ be a nonzero map with $C \in \underline{C}$. We replace C by its image C' which is also in $\underline{C}$ since $\underline{C}$ is a Serre subcategory. $C' \neq 0$ since $f \neq 0$. Therefore, $C' \cap M = C'' \neq 0$ since A is an essential extension. But $C'' \in \underline{C}$ since $\underline{C}$ is a Serre subcategory. Therefore, we have a nonzero map $C'' \longrightarrow M$ which is impossible. Thus, $\underline{M}$ is closed under essential extensions.

Now we show we can recover $\underline{C}$ from $\underline{M}$. Pick $A \in \underline{A}$ with $(A, M) = 0$ for all $M \in \underline{M}$. We want $A \in \underline{C}$. There exists a $\underline{C}$ envelope $u_A: A \longrightarrow RA$ where $RA \in \underline{L}$. $u_A = 0$ since $(A, M) = 0$ for all $M \in \underline{M}$ and $\underline{M}$ contains $\underline{L}$. But u_A is a $\underline{C}$ iso. Therefore $A = \ker u_A \in \underline{C}$.

Now for the other direction. We are given $\underline{M}$ and construct $\underline{C} = \{C \mid (C, M) = 0 \text{ for all } M \in \underline{M}\}$.

1) $\underline{C}$ is a Serre subcategory.

Lemma 1.17. $C \in \underline{C}$ if and only if $(C, I) = 0$ for all injectives $I \in \underline{M}$.

Proof. $\Longrightarrow$ trivial.

$\Longleftarrow$ Let $M \in \underline{M}$. Then there exists an injective envelope, $0 \longrightarrow M \longrightarrow I$. $I \in \underline{M}$ since M is closed under essential extensions.

$0 \to M \to I$ exact implies

$0 \to (C, M) \to (C, I)$ is exact. Hence, $(C, M) = 0$ if $(C, I) = 0$. Done.

Let $0 \to C' \to C \to C'' \to 0$ be exact in $\underline{A}$. Let I be any injective in M. Then

$0 \leftarrow (C', I) \leftarrow (C, I) \leftarrow (C'', I) \leftarrow 0$ is exact and $\underline{C}$ is clearly a Serre subcategory.

2) $\underline{C}$ is localizing.

<u>Remark</u>. If $\underline{A}$ was right complete, this would be immediate because if $C_\alpha \in \underline{C}$, $(\coprod C_\alpha, M) = \prod (C_\alpha, M) = 0$ and $\coprod C_\alpha \in \underline{C}$.

Let $A \in \underline{A}$. Consider all quotient objects M_α of A with $M_\alpha \in \underline{M}$. Since $\underline{A}$ is well powered this is a set. $\prod M_\alpha \in \underline{M}$. Consider the exact sequence

$$* \qquad 0 \to C \xrightarrow{i} A \to \prod M_\alpha \quad .$$

Let f be any map $f: A \to M$ with $M \in \underline{M}$. Then f annihilates C since $\underline{M}$ closed under subobjects implies f factors through $\prod M_\alpha$. We claim $C \in \underline{C}$. Let I be an injective in $\underline{M}$. Let $g: C \to I$. It is enough to show $g = 0$. Consider the diagram

$$\begin{array}{ccccc} 0 & \to & C & \xrightarrow{i} & A \\ & & {\scriptstyle g}\downarrow & \swarrow {\scriptstyle f} & \\ & & I & & \end{array} \quad .$$

f exists since I is injective. $fi = 0$ since all maps from A to an object of M annihilate C. Therefore, $g = 0$ and $C \in \underline{C}$.

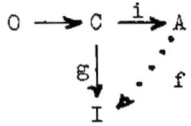

C is even the largest $\underline{C}$ subobject of A. Let $0 \longrightarrow D \longrightarrow A$ be exact with $D \in \underline{C}$. Then $D \longrightarrow A \longrightarrow \prod M_\alpha$ is the zero map and, hence, factors through C. Hence $\underline{C}$ is localizing by corollary 1.8.

3) $\underline{C}$ gives back $\underline{M}$. Let $A \in \underline{A}$ such that $(C, A) = 0$ for all $C \in \underline{C}$. Construct $\prod M_\alpha$ where M_α runs over all $\underline{M}$ quotient objects of A. We have $0 \longrightarrow C \longrightarrow A \longrightarrow \prod M_\alpha$ as before. Since $C \in \underline{C}$ and $(C, A) = 0$ $C = 0$. But $\underline{M}$ is closed under subobjects, and, hence, $A \in \underline{M}$. Done.

<u>Definition</u>. Let $\underline{A}$ and $\underline{B}$ be categories. Then $\underline{A}$ and $\underline{B}$ are <u>equivalent</u> if there exist covariant functors $S: \underline{A} \longrightarrow \underline{B}$ and $T: \underline{B} \longrightarrow \underline{A}$ such that TS is naturally equivalent to id $\underline{A}$ and ST is naturally equivalent to id $\underline{B}$.

<u>Proposition 1.18</u>. Let $S: \underline{A} \longrightarrow \underline{B}$ be a covariant functor. Then S is an equivalence if and only if 1) S is fully faithful, and 2) for each $B \in \underline{B}$ there exists an $A \in \underline{A}$ and an iso in $\underline{B}$ with $B \cong S(A)$.

<u>Proof</u>. $\Longrightarrow$ Let f and $g: A \longrightarrow B$ such that $S(f) = S(g)$. Then $(TS)(f) = (TS)(g)$ and the following two squares commute:

$$\begin{array}{ccc} A & \longrightarrow & TS(A) \\ g\downarrow & & \downarrow TS(g) \\ B & \longrightarrow & TS(B) \end{array} \quad = \quad \begin{array}{ccc} TS(A) & \longleftarrow & A \\ TS(f)\downarrow & & \downarrow f \\ TS(B) & \longleftarrow & B \end{array}$$

where all horizontal maps are isomorphisms. Thus $f = g$ and S is faithful.

Now we show that S is fully faithful. Let $h: SA \longrightarrow SB$ be a map in $\underline{B}$. Then $Th: TSA \longrightarrow TSB$ and the following diagram commutes:

$$\begin{array}{ccc} A & \xrightarrow{f} & B \\ \Big| S & & \Big| S \\ TSA & \xrightarrow[Th]{} & TSB \end{array}$$ where f is the

map induced by $A \longrightarrow TSA \xrightarrow[Th]{} TSB \longrightarrow B$. Then $Th = TS(f)$. But T is faithful so $h = S(f)$ or S is fully faithful.

B is isomorphic to ST(B) so A = T(B) takes care of the second condition.

$\Longleftarrow$ For each $B \in \underline{B}$ choose an $A \in \underline{A}$ such that SA is isomorphic to B. Fix an iso η_B: $SA \longrightarrow B$. Set $T(B) = A$. Then η_B is an iso from STB to B.

We now must say what T does to maps. Let $f: B \longrightarrow B'$. This induces a map $h: ST(B) \longrightarrow ST(B')$ so that the following square commutes:

$$\begin{array}{ccc} B & \xrightarrow{f} & B' \\ \eta_B \Big\downarrow & & \Big\downarrow \eta_{B'} \\ ST(B) & \longrightarrow & ST(B') \end{array} .$$

S is fully faithful so there exists a unique $g: TB \longrightarrow TB'$ so that $h = S(g)$. Define $T(f) = g$.

With this definition of T it is easy to see that T is a functor and that S and T are equivalences. Done.

Let $\underline{A}$ and $\underline{B}$ be equivalent categories with $S: \underline{A} \longrightarrow \underline{B}$ and $T: \underline{B} \longrightarrow \underline{A}$ a pair of equivalences.

1) S is a left and right adjoint of T.

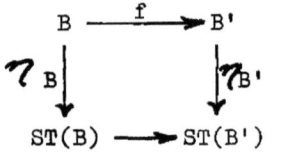

$(SA, B) \xrightarrow{\cong} (TSA, TB) \xrightarrow{\cong} (A, TB)$ shows that S is a right adjoint of T. The other follows by symmetry. Thus, S and T preserve $\varinjlim$ and $\varprojlim$ when defined.

2) If $\underline{A}$ is abelian then so is $\underline{B}$. (B, B') is isomorphic to (TB, TB') which is an abelian group. Use the isomorphism to pull the group structure back. Composition in $\underline{B}$ is still bilinear since T is a functor. B is a 0 object in $\underline{B}$ if and only if TB = 0 if and only if $1_B = 0$. Coim is isomorphic to image by applying T and pulling back the isomorphism by S.

3) If $A \in \underline{A}$ is $\left\{\begin{array}{l}\text{projective}\\ \text{injective}\\ \text{a generator,}\\ \text{a cogenerator}\end{array}\right.$ then so is SA in $\underline{B}$.

4) Let $\underline{C}$ be another category and $F: \underline{C} \longrightarrow \underline{A}$ a functor. Then

F is $\left\{\begin{array}{l}\text{left exact}\\ \text{right exact}\\ \text{faithful}\\ \text{full}\end{array}\right.$ if and only if SF is.

We now examine what equivalent categories look like. Let $\underline{C}$ be any class and S: $\underline{C} \longrightarrow$ objects of $\underline{A}$ where $\underline{A}$ is any category. Assume for every $A \in \underline{A}$ there exists a $C \in \underline{C}$ and an iso, f, in $\underline{A}$ with $S(C) \xrightarrow{f} A$. Define (C, C') = (S(C), S(C')) for any pairs C and $C' \in \underline{C}$. Composition is defined by the following diagram.

$$\begin{array}{ccccc}(C, C') & \times & (C', C'') & \longrightarrow & (C, C'')\\ \| & & \| & & \| \\ (SC, SC') & \times & (SC', SC'') & \longrightarrow & (SC, SC'').\end{array}$$

That is by pulling back the composition in $\underline{A}$. This clearly makes $\underline{C}$ into a category which is equivalent to $\underline{A}$ by S.

Now we claim that if a category B is equivalent to A then B looks like one of the above. Let S: B $\longrightarrow$ A be an equivalence and C be the objects of B. Then for each A $\in$ A there is a C $\in$ C and an iso f: S(C) $\longrightarrow$ A. Thus we have S: C $\longrightarrow$ objects of A. The previous construction clearly recovers B.

Examples. 1) Skeletal subcategory. Choose one object in each isomorphism class of A. Let C be the class of chosen objects. For S we take the inclusion of C in objects of A and make C equivalent to A. The category so constructed has one object in each isomorphism class.

2) Replete category. Let U be the universal class and let C = U x objects of A. We map C to A by projection onto second factor. Then we make C into a category such that each class of C is a proper class.

Next we examine when a category is equivalent to a category of modules and homomorphisms.

Definition. Let A be an abelian category and P a projective generator of A. Then P is small if (P,) preserves direct sums.

Remark. Bass in his Oregon Morita theorem notes defines progenerator to be a small projective generator.

Theorem 1.19. Let A be an abelian category. Then A is equivalent to M_R, the category of right R modules over some ring R, if and only if A is right complete and has a small projective generator.

Proof. $\Longrightarrow$ R is a small projective generator for M_R since (R, M) $\cong$ M. M_R is right complete.

$\Longleftarrow$ Let P be a small projective generator of a right complete abelian category $\underline{A}$. Let R = (P, P). We claim $\underline{A} \xrightarrow{(P,\ )} \underline{M}_R$ is an equivalence. To show this we must demonstrate the following:

1) (P,) is faithful,

2) If M is an R module then M is isomorphic to (P, A) for some $A \in \underline{A}$, and

3) (P,) is full.

1) follows easily since P is a generator.

2) We start with free modules.

a) Let F be a free R module. Then

$F = \sum_{a \in J} R$. $F \cong (P, \sum_{a \in J} P) = \sum_{a \in J} (P, P) = \sum_{a \in J} R = F$

since P is small.

b) To finish 2) we need that (P,) = T( ) is full when the images are free modules. Let $A = \sum_{a \in I} P$ and $B = \sum_{b \in J} P$. We want that (A, B) is isomorphic to (TA, TB).

T preserves direct sums implies

$$\begin{array}{ccc} (\sum_{a \in I} P, \sum_{b \in J} P) & \longrightarrow & (\sum_{a \in I} TP, \sum_{b \in J} TP) \\ \wr\| & & \wr\| \\ \prod_{a \in I} (P, \sum_{b \in J} P) & \longrightarrow & \prod_{a \in I} (TP, \sum_{b \in J} TP). \end{array}$$

Hence it is enough to show that

$$\begin{array}{ccc} (P, \sum_{b \in J} P) & \longrightarrow & (TP, \sum_{b \in J} TP) \\ \| & & \| \\ \sum_{b \in J} (P, P) & \longrightarrow & \sum_{b \in J} (TP, TP) \end{array}$$ is onto.

But we are in $\underline{M}_R$ so if each component is onto then the sum is onto. We check that the map

$$\begin{array}{c} R = (P, P) \longrightarrow (TP, TP) = R \\ \quad\quad\quad\quad\quad\quad\quad \| \quad\ \| \\ \quad\quad\quad\quad\quad\quad\quad R \quad R \end{array}$$

is the identity map. Pick $r \in R = (R, P)$. Then $r: P \longrightarrow P$ and $T(r): TP \longrightarrow TP$ is given by

$$\begin{array}{ccc} T(P) & \longrightarrow & T(P) \\ \| & & \| \\ (P, P) & \longrightarrow & (P, P) \\ a & \rightsquigarrow & ra \end{array} .$$

This shows that $(P, P) \longrightarrow (TP, TP)$ coincides with

$R \longrightarrow (R, R)$ given by

$r \rightsquigarrow f_r: a \longrightarrow ra$ so the map is identity map if we identify $(R, R) = R$, and b) is done.

Let $M \in \underline{M}_R$. We want M to be isomorphic to $T(C)$ for some $C \in \underline{A}$. Let

$F' \xrightarrow{f} F \longrightarrow M \longrightarrow 0$ be a free

resolution of M. Then by a) there exist $A = \sum P$ and $B = \sum P$ such that the following diagram commutes:

$$\begin{array}{ccc} F' & \xrightarrow{f} & F \\ \downarrow S & & \downarrow S \\ TA & \xrightarrow{f'} & TB \end{array} .$$

M is isomorphic to coker f'. f' = T(g) for some g: A $\longrightarrow$ B by b). Let C be the coker of g. Then T(C) is the coker of T(g) = f' since T is exact. Thus M = coker f is isomorphic to T(C) for some C $\in$ $\underline{A}$, and 2) is proved.

3) It remains to show that T is full, that is that (A, B) $\longrightarrow$ (TA, TB) is onto. Let f: TA $\longrightarrow$ TB. Construct resolutions of A and B

$$X' \longrightarrow X \longrightarrow A \longrightarrow 0 \quad \text{and}$$

$$Y' \longrightarrow Y \longrightarrow B \longrightarrow 0 \quad \text{where } X, X', Y, Y' \text{ are all}$$

direct sums of copies of P. Applying the exact functor T we get

$$* \qquad \begin{array}{ccccccc} TX' & \longrightarrow & TX & \longrightarrow & TA & \longrightarrow & 0 \\ \downarrow f'' & & \downarrow f' & & \downarrow f & & \\ TY' & \longrightarrow & TY & \longrightarrow & TB & \longrightarrow & 0 \end{array} .$$

f' and f" exist by the usual method of extending a map to a map on projective resolutions.

But T is full on objects that become free so there exist h', h", and g with T(h") = f", T(h') = f'. The following diagram commutes because T is faithful.

$$\begin{array}{ccccccc} X' & \longrightarrow & X & \longrightarrow & A & \longrightarrow & 0 \\ h'' \downarrow & & h' \downarrow & & g \vdots & & \\ Y' & \longrightarrow & Y & \longrightarrow & B & \longrightarrow & 0 \end{array}$$

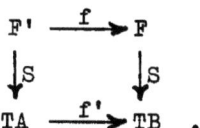

where g is the unique map making the diagram commute.

Applying T again we obtain that T(g) and f both make * commute. But TX $\longrightarrow$ TA is an epi. Therefore T(g) = f and T is full.

Thus we have an equivalence by proposition 1.18. Done.

Example. Let R be a ring then $R \oplus \ldots \oplus R$ (n times) is a small projective generator. Let $M_n(R)$ be the ring of n x n matrices over R. Then we have shown that $\underline{M}_R$ and $\underline{M}_{M_n(R)}$ are equivalent categories.

Theorem 1.20. (Watts). Let A and B be rings and $F: {}_A\underline{M} \longrightarrow {}_B\underline{M}$ be a covariant functor. Then the following are equivalent:

1) F has a right adjoint,

2) F is right exact and preserves direct sums, and

3) There exists an A-B bimodule M such that F is naturally isomorphic to $M \otimes_A \_\_$.

Proof. 1) $\Longrightarrow$ 2) is trivial.

3) $\Longrightarrow$ 1) $\mathrm{Hom}_B(M \otimes_A X, Y) = \mathrm{Hom}_A(X, \mathrm{Hom}_B(M, Y))$ gives the right adjoint.

2) $\Longrightarrow$ 3) Let M = F(A) and $a \in A$. Then we define $\hat{a}: A \longrightarrow A$ by $\hat{a}(x) = xa$. This is a map of left A-modules. Given $a, b \in A$

$$\widehat{ab} = \hat{b}\hat{a}.$$

The action of A and B on M = F(A) commute since $F(\hat{a})$ is a B map. Thus M is an A-B bimodule. Let $X \in {}_A\underline{M}$. Define for each $x \in X$ $\hat{x}: A \longrightarrow X$ by $\hat{x}(a) = ax$. Then define $\eta_X: M \otimes_A X \longrightarrow F(X)$

by $\eta_X(m \otimes x) = F(\hat{x})m$ where $x \in X$ and $m \in M$.

We need to check $\eta_X(ma \otimes x) = \eta_X(m \otimes ax)$.

$\eta_X(ma \otimes x) = F(\hat{x})F(\hat{a})m$ and

$\eta_X(m \otimes ax) = F(\widehat{ax})m = F(\hat{x})F(\hat{a})m$.

Thus we have a natural transformation of functors. We want it to be a natural isomorphism.

<u>Case 1</u>. $X = A$. Then we have $M \otimes_A A \longrightarrow F(A) = M$ and the induced map is the identity.

<u>Case 2</u>. X is a free A module. Both functors preserve direct sums so η_X is an isomorphism.

<u>Case 3</u>. General case. Let

$X_1 \longrightarrow X_0 \longrightarrow X \longrightarrow 0$ be a free resolution of X. F is right exact. Hence we get the diagram

$$\begin{array}{ccccccc} M \otimes_A X_1 & \longrightarrow & M \otimes_A X_0 & \longrightarrow & M \otimes_A X & \longrightarrow & 0 \\ \downarrow{\scriptstyle \eta_{X_1}} & & \downarrow{\scriptstyle \eta_{X_2}} & & \downarrow{\scriptstyle \eta_X} & & \\ FX_1 & \longrightarrow & FX_0 & \longrightarrow & FX & \longrightarrow & 0 \end{array}$$

where the rows are exact and η_{X_1} and η_{X_2} are isomorphisms. Hence F and $M \otimes_A$ are naturally isomorphic. Done.

Let A be a ring, $\underline{C}$ be a localizing Serre subcategory of ${}_A\underline{M}$, $\underline{L}$ be the full subcategory of $\underline{C}$ closed objects, and $R\colon {}_A\underline{M} \longrightarrow \underline{L}$ be the reflection. We have

$$\begin{array}{ccccc} (A, A) & \xrightarrow{R} & (RA, RA) & \xrightarrow[\approx]{(u_A, 1)} & (A, RA) \\ \| & & & & \| \\ A^0 & & \longrightarrow & & RA^0 \end{array}$$

where R is a ring endomorphism and the isomorphism holds since u_A is a $\underline{C}$ envelope.

In the above A becomes A°, the opposite ring, so

we make RA into a ring by lifting the action of (RA, RA)°. So RA is a ring and $A \longrightarrow RA$ is a ring homomorphism. Let $L \in \underline{L}$. Then (RA, L) is a right (RA, RA) module. Therefore, (A, L) is a left (RA, RA)° module or L is a left RA module. So each $L \in \underline{L}$ is a left RA module. Therefore we have a functor $\underline{L} \longrightarrow {}_{RA}\underline{M}$ via $L \rightsquigarrow (RA, L) = L$ as an RA module. This is full and faithful because $(L_1, L_2)_{\underline{L}} = (L_1, L_2)_{\underline{M}_A} = \mathrm{Hom}_A(L_1, L_2) \subset \mathrm{Hom}_{RA}(L_1, L_2) \subset (L_1, L_2)_{\underline{L}}$. It can fail to be exact.

Theorem 1.21. $\underline{L} \longrightarrow {}_{RA}\underline{M}$ is an equivalence if and only if $i: \underline{L} \longrightarrow {}_A M$ is exact and preserves direct sums. If so, then the composition ${}_A\underline{M} \longrightarrow \underline{L} \longrightarrow {}_{RA}\underline{M}$ is naturally equivalent to $RA \otimes_A \_\_$ and RA is a flat right A module.

Proof. $\Longrightarrow$ Let $\underline{L} \longrightarrow {}_{RA}\underline{M}$ be an equivalence. R is exact and preserves $\varinjlim$. Therefore so is ${}_A\underline{M} \longrightarrow {}_{RA}\underline{M}$. By Watt's theorem this is naturally isomorphic to $RA \otimes_A \_\_$. Since it is exact, RA is flat as a right A-module. Let i be the right adjoint of R. Then i is equivalent to the right adjoint of $RA \otimes_A \_\_$. But the right adjoint of $RA \otimes_A \_\_$ is $\mathrm{Hom}_{RA}(RA, \ )$.

$$\mathrm{Hom}_{RA}(RA, \_\_): {}_{RA}\underline{M} \longrightarrow {}_A\underline{M}$$

$$M \rightsquigarrow M \text{ as an A module.}$$

This is clearly exact and preserves direct sums. Hence so does i.

$\Longleftarrow$ Hom(RA,): $\underline{L} \longrightarrow {}_{RA}\underline{M}$ is full and faithful. But $(RA, L)_{\underline{L}} = (A, iL)_{{}_A\underline{M}}$. But i is exact and preserves direct sums and so does (A, __). Hence so does the composition. This shows that RA is a small projective generator. Thus $\underline{L} \longrightarrow {}_{RA}\underline{M}$ is an equivalence by theorem 1.19.

Remark. Since i: $\underline{L} \longrightarrow {}_A\underline{M}$ is just the inclusion functor and is always left exact, the conditions that i be exact and preserve direct sums are clearly equivalent to the conditions that $\underline{L}$ be closed under the formation of cokernels and direct sums in ${}_A\underline{M}$. e.g., i is exact if and only if $0 \longrightarrow M' \longrightarrow M \longrightarrow M'' \longrightarrow 0$ in ${}_A\underline{M}$ and M', M $\in \underline{L}$ imply M" $\in \underline{L}$. Similarly i preserves direct sums if and only if $M = \coprod M_\alpha$ in ${}_A\underline{M}$, all $M_\alpha \in \underline{L}$ implies $M \in \underline{L}$.

We now illustrate the above results by reconstructing the classical theory of rings of quotients following Gabriel.

Let R be a ring and S a multiplicatively closed subset of R, i.e., s, t $\in$ S imply st $\in$ S. We assume 1 $\in$ S for convenience. Let $\underline{C}$ be the full subcategory of ${}_R\underline{M}$ of all modules C such that for each c $\in$ C, there exists an s $\in$ S with sc = 0. It is trivial to verify that $\underline{C}$ is a Serre subcategory and is closed under directed $\varinjlim$. Therefore $\underline{C}$ is localizing. In order to identify $\underline{L}$ and apply theorem 2.15 we must impose some extra conditions on S. We choose the following two classical conditions both of which are trivially true if S is central in R.

(I) If a $\in$ R, s $\in$ S, there exist b $\in$ R, t $\in$ S so that ta = bs.

(II) If r $\in$ R, s $\in$ S and rs = 0, there exists t $\in$ S such that tr = 0.

Proposition 1.22. *If* S *satisfies* (I) *and* (II) *and* $M \in {}_R\underline{M}$, *then*

(a) $M \in \underline{M}$ *if and only if* $M \xrightarrow{\hat{s}} M$ (given by $x \rightsquigarrow sx$) *is a monomorphism for all* $s \in S$.

(b) $M \in \underline{L}$ *if and only if* $M \xrightarrow{\hat{s}} M$ *is an isomorphism for all* $s \in S$.

Note. $M \xrightarrow{\hat{s}} M$ will not be an R-homomorphism if s is not central in R.

Proof:

Let $C(M) = \{x \in M | sx = 0 \text{ for some } s \in S\}$. Clearly every $\underline{C}$-subobject of M is contained in C(M). If we can show that C(M) is itself an R-submodule it will lie in $\underline{C}$ and be the largest $\underline{C}$-submodule of M. Let $x \in C(M)$ and $a \in R$. By definition, there is an $s \in S$ with $sx = 0$. By I we find $ta = bs$ with $t \in S$. Therefore $tax = bsx = 0$ and $ax \in C(M)$. Suppose now $x, y \in C(M)$. We must show $x + y \in C(M)$. Let $s \in S$ with $sx = 0$. Then $s(x + y) = sy$. Since $y \in C(M)$, $sy \in C(M)$ as we have seen. Therefore there is some $t \in S$ with $tsy = 0$ so $ts(x + y) = 0$ and $ts \in S$ which is multiplicatively closed.

Now $M \in \underline{M}$ if and only if the largest $\underline{C}$-subobject C(M) is 0, i.e., if and only if $sx = 0$, $s \in S$, $x \in M$ implies $x = 0$. This is (a).

For (b) we observe that by (a) each of the conditions $M \in \underline{L}$, $M \xrightarrow{\hat{s}} M$ all $s \in S$ implies $M \in \underline{M}$. Therefore we can assume $M \in \underline{M}$. Let I be the injective envelope of M, $0 \to M \to I \to X \to 0$. By proposition 1.6. , $M \in \underline{L}$ if and only if $X \in \underline{M}$ and also $I \in \underline{M}$.

Consider the diagram

$$\begin{array}{ccccccccc} 0 & \longrightarrow & M & \longrightarrow & I & \longrightarrow & X & \longrightarrow & 0 \\ & & \downarrow{\hat{s}} & & \downarrow{\hat{s}} & & \downarrow{\hat{s}} & & \\ 0 & \longrightarrow & M & \longrightarrow & I & \longrightarrow & X & \longrightarrow & 0 \end{array} .$$

We know $M \in \underline{L} \Longleftrightarrow X \in \underline{M} \Longleftrightarrow X \xrightarrow{\hat{s}} X$ is a monomorphism for all $s \in S$. If $I \xrightarrow{\hat{s}} I$ is an isomorphism for all $s \in S$, the five lemma shows that $X \xrightarrow{\hat{s}} X$ is a monomorphism if and only if $M \xrightarrow{\hat{s}} M$ is an isomorphism and we will be done.

<u>Lemma 1.23</u>. <u>If</u> $I \in \underline{M}$ <u>and</u> I <u>is injective then</u> $I \xrightarrow{\hat{s}} I$ <u>is an isomorphism for all</u> $s \in S$, <u>assuming that</u> S <u>satisfies</u> (I) <u>and</u> (II).

<u>Proof</u>.

Since $I \in \underline{M}$, $I \xrightarrow{\hat{s}} I$ is a monomorphism by proposition 2.16 (a) using (I). We must show $I \xrightarrow{\hat{s}} I$ is onto. Let $x \in I$. Define an R-homomorphism $Rs \xrightarrow{f} I$ by $f(rs) = rx$. To see that f is well defined, it will suffice to show that $rs = 0$ implies $rx = 0$. By (II), there is some $t \in S$ such that $tr = 0$. Therefore $trx = 0$ but $I \xrightarrow{\hat{t}} I$ is a monomorphism for all $t \in S$ so $rx = 0$. Since I is injective, f extends to an R-homomorphism $g: R \longrightarrow I$ and $x = f(s) = g(s) = sg(1) \in sI$.

<u>Corollary 1.24</u>. <u>If</u> S <u>satisfies</u> (I) <u>and</u> (II), <u>then</u> L <u>is closed under cokernels and direct sums in</u> ${}_A\underline{M}$. <u>Therefore theorem</u> 1.21 <u>applies</u>.

<u>Proof</u>.

This is immediate from proposition 1.22. b). I.e., if $L_i \in \underline{L}$, then for $s \in S$, $L_i \xrightarrow{\hat{s}} L_i$ is an isomorphism. Therefore so is $\coprod L_i \xrightarrow{\hat{s}} \coprod L_i$. If $0 \to A' \to A \to A'' \to 0$ and $A', A \in \underline{L}$, then the 5 lemma on

$$\begin{array}{ccccccccc} 0 & \to & A' & \to & A & \to & A'' & \to & 0 \\ & & \downarrow{\scriptstyle \hat{s}} & & \downarrow{\scriptstyle \hat{s}} & & \downarrow{\scriptstyle \hat{s}} & & \\ 0 & \to & A' & \to & A & \to & A'' & \to & 0 \end{array}$$

shows that $A'' \xrightarrow{\hat{s}} A''$ is an isomorphism.

<u>Notation</u>. We denote the reflection $R: {}_A\underline{M} \to \underline{L}$ by $A \rightsquigarrow {}_SA$ and identify $\underline{L}$ with ${}_{{}_SR}\underline{M}$.

By theorem 1.21, ${}_SR$ is flat as a right R-module and ${}_SA \approx {}_SR \otimes_R A$ naturally. It is not clear whether ${}_SR$ is flat as a left R-module.

<u>Remark</u>. If S satisfies

(I') If $a \in R$, $s \in S$, there exist $b \in R$, $t \in S$ so that $at = sb$.

(II') If $r \in R$, $s \in S$, and $sr = 0$, there exists $t \in S$ such that $rt = 0$,

then the above considerations apply to <u>right</u> modules $\underline{M}_R$ and we get $A \to A_S$, $R \to R_S$ and R_S is flat as a left R-module.

We can easily recover the classical construction for ${}_SA$. Consider the $\underline{C}$ isomorphism $u_A: A \to {}_SA$. Since ${}_SA \xrightarrow{\hat{s}} {}_SA$ is an isomorphism for $s \in S$, we can define $\hat{s}^{-1}$ as the inverse of this.

<u>Proposition 1.25</u>. Every element of ${}_SA$ has the form $s^{-1}u_A(x)$ for $x \in A$, $s \in S$. We have $s^{-1}u_A(x) = t^{-1}u_A(y)$ if and only if there exist $r \in R$, $w \in S$ such that $wt = rs$ and $rx = wy$.

<u>Proof</u>.

Let $B = \{s^{-1}u_A(x) | s \in S, x \in A\}$. Then B is an R-submodule of ${}_SA$. In fact if $r \in R$, (I) gives $t \in S$, $b \in R$ so $tr = bs$. Therefore $trs^{-1}u_A(x) = bss^{-1}u_A(x) = u_A(bx)$ so $rs^{-1}u_A(x) = t^{-1}u_A(bx) \in B$. Now consider $s^{-1}u_A(x) + t^{-1}u_A(y)$. By (I) we find $ws = ct$ with $w \in S$. Then $s^{-1}u_A(x) + t^{-1}u_A(y) = s^{-1}[u_A(x) + st^{-1}u_A(y)] = s^{-1}w^{-1}[wu_A(x) + wst^{-1}u_A(y)] = s^{-1}w^{-1}u_A(wx + cy)$.

Next we note $B \in \underline{L}$. Clearly $B \xrightarrow{\hat{s}} B$ is a monomorphism since $B \subset S^{-1}A$. It is onto because $t^{-1}u_A(x) = s(ts)^{-1}u_A(x)$.

Since $B \in \underline{L}$, $B \subset {}_SA$, and $u_A\colon A \longrightarrow B$, we see that B is a $\underline{C}$ envelope of A so $B = {}_SA$.

Now suppose $s^{-1}u_A(x) = t^{-1}u_A(y)$. Find $w_1 \in S, r_1 \in R$ so $w_1t = r_1s$. Then $w_1u_A(y) = w_1tt^{-1}u_A(y) = w_1ts^{-1}u_A(x) = r_1u_A(x)$ so $w_1y - r_1x \in \ker u_A \in \underline{C}$. By the definition of $\underline{C}$, there is some $v \in S$ so that $v(w_1y - r_1x) = 0$. Set $w = vw_1$ and $r = ur_1$. Conversely, if $w \in S$, $wt = rs$, and $rx = wy$, then $wt(s^{-1}u_A(x) - t^{-1}u_A(y)) = $
$= ru_A(x) - wu_A(y) = u_A(rx - wy) = 0$. Hence, $s^{-1}u_A(x) = t^{-1}u_A(y)$.

Done.

It follows from proposition 1.25 that we can construct ${}_SA$ by considering pairs $(s, x) \in S \times A$ and defining an equivalence relation by $(s, x) \sim (t, y)$ if and only if there are $r \in R$ and $w \in S$ such that $wt = rs$ and $rx = wy$.

If S is central in R, this reduces to the usual equivalence relation $(s, x) \sim (t, y)$ if and only if there is a $v \in S$ with $v(tx - sy) = 0$. Here, of course, $r = vt$ and $w = vs$. Conversely, if wt = rs and rx = wy, then srx = swy and w(tx - sy) = 0. Hence, in this case, ${}_SR = R_S$.

SECTION II

QUOTIENT CATEGORIES

The purpose of this section is to generalize the notation of localization. There are two problems we shall investigate. Let A be an abelian category and C a Serre subcategory.

Problem 1. Find an exact functor T and an abelian category A/C with T: A ⟶ A/C such that if F: A ⟶ B is an exact functor which annihilates C then there exists a unique functor G with F = GT. G will be exact. This is possible if A is well powered or if C is localizing.

Problem 2. Find an exact functor T: A ⟶ D which is universal for exact functors which annihilate C up to natural isomorphism. That is given the diagram

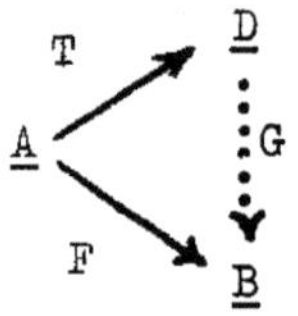

where F is exact and annihilates C. Then G: D ⟶ B is exact and GT is naturally isomorphic to F and any other G' is naturally isomorphic to G.

If C is a localizing subcategory, then the reflection R: A ⟶ L into the full subcategory of C closed objects solves problem 2. Let $F: \underline{A} \longrightarrow \underline{B}$ be exact and annihilate $\underline{C}$. Then we have the diagram

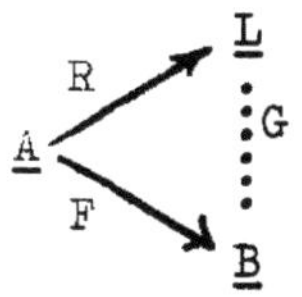

.

Putting $G = F|_{\underline{L}}$ always works. We need to show that $F|_{\underline{L}}$ is exact. Let

$$0 \longrightarrow L' \longrightarrow L \longrightarrow L'' \longrightarrow 0 \quad \text{be exact in } \underline{L}.$$

Then $0 \longrightarrow L' \longrightarrow L \longrightarrow L'' \longrightarrow Q \longrightarrow 0$ is exact in $\underline{A}$. Applying R gives

$$0 \longrightarrow L' \longrightarrow L \longrightarrow L'' \longrightarrow RQ \longrightarrow 0$$

exact in $\underline{L}$ or $RQ = 0$. Hence $Q \in \underline{C}$. Then

$$0 \longrightarrow FL' \longrightarrow FL \longrightarrow FL'' \longrightarrow FQ \longrightarrow 0 \text{ is exact in}$$

$\underline{B}$. But F annihilates $\underline{C}$ so $FQ = 0$ and $F|_{\underline{L}}$ is exact.

Next we show GR is naturally isomorphic to F. For each $A \in \underline{A}$ we have $u_A: A \longrightarrow RA$ where the ker and coker of u_A are in $\underline{C}$. Therefore, $F(u_A)$ is an iso. But, $FRA = GRA$ since $G = F|_{\underline{L}}$, and $F(A) \longrightarrow FR(A)$ is a natural iso. Therefore $F(A) \longrightarrow GR(A)$ is a natural iso.

Finally suppose $\eta: GR \longrightarrow F$ is a natural isomorphism. We want to show G is naturally isomorphic to $F|_{\underline{L}}$.

$$\eta|_{\underline{L}}: GR|_{\underline{L}} \longrightarrow F|_{\underline{L}} \text{ isomorphically.}$$

But R is the identity on $\underline{L}$ so $GR|_{\underline{L}} = G$ and G is naturally isomorphic to $F|_{\underline{L}}$.

It could happen that $\underline{A}$ was very big and $\underline{L}$ small (by creating a new equivalent category with lots of objects in each is isomorphism type of $\underline{A}$ not in $\underline{L}$). Then the above construction won't work for problem 1 because R will identify too many objects to be able to factor F as GR. This is the only difficulty. We remove it by the following construction.

We have obj $\underline{A} \xrightarrow[R]{} \underline{L}$. Pullback the structure of $\underline{L}$ via R to obtain a new category $\underline{A}/\underline{C}$ equivalent to $\underline{L}$ but with obj $\underline{A}/\underline{C}$ = obj $\underline{A}$ and

$$(A, B)_{\underline{A}/\underline{C}} = (RA, RB)_{\underline{L}} = (RA, RB)_{\underline{A}}.$$

We have the diagram

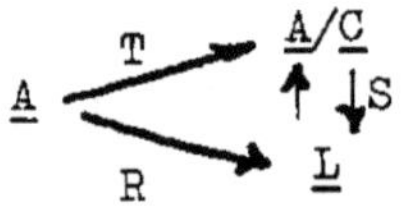

where S is an equivalence, $T(A) = A$, and $T: (A, B)_{\underline{A}} \longrightarrow (A, B)_{\underline{A}/\underline{C}} = (RA, RB)_{\underline{A}}$. T is a functor, and $ST = R$. $\underline{A}/\underline{C}$ is abelian since S is an equivalence. T is exact and annihilates $\underline{C}$ since $T(A) = 0$ if and only if $STA = 0 = RA$. But T cannot factor through $\underline{L}$ since no objects are identified. $T: \underline{A} \longrightarrow \underline{A}/\underline{C}$ is universal for exact functors on $\underline{A}$ which annihilate $\underline{C}$. Let $F: \underline{A} \longrightarrow \underline{B}$ be exact and annihilate $\underline{C}$. Then we will show that there exists a unique G such that $GT = F$. If G exists, it will be exact since $\underline{A} \xrightarrow{R} \underline{L}$ is universal up to natural isomorphism, and, therefore, so is T. But exactness is preserved under equivalence. If G exists, $G = F$ on objects since

T = identity on objects. We now check what G must do to maps. Let $f \in (A, B)_{\underline{A}/\underline{C}} = (RA, RB)_{\underline{A}}$. Then in A we have

$$
* \qquad \begin{array}{ccc} A & & B \\ {\scriptstyle u_A}\downarrow & & \downarrow \\ RA & \xrightarrow[f]{} & RB \end{array} .
$$

Applying T gives

$$
** \qquad \begin{array}{ccc} TA & \xrightarrow{f} & TB \\ \downarrow & & \downarrow{\scriptstyle T(u_B)} \\ TRA & \xrightarrow[T(f)]{} & TRB \end{array} \qquad \text{in } \underline{A}/\underline{C} .
$$

$T(u_A)$ and $T(u_B)$ are isos since u_A and u_B are $\underline{C}$ isos and T annihilates $\underline{C}$. Apply R to * yields

$$
\begin{array}{ccc} RA & \xrightarrow{f} & RB \\ {\scriptstyle R(u_B)}\downarrow & & \downarrow{\scriptstyle R(u_A)} \\ RRA & \longrightarrow & RRB \end{array} .
$$

But RA and RB $\in \underline{L}$ and R is the identity on $\underline{L}$. Hence, RRA = RA, RRB = RB, R(f) = f, $R(u_A) = 1_{RA}$, $R(u_B) = 1_{RB}$.

We assume we have G and apply it to ** giving

$$
\begin{array}{ccc} GTA & \xrightarrow{G(f)} & GTB \\ {\scriptstyle GT(u_A)}\downarrow & & \downarrow{\scriptstyle GT(u_B)} \\ GTRA & \xrightarrow[GT(f)]{} & GTRB \end{array} .
$$

But GT = F so the above really is

$$
\begin{array}{ccc} FA & \xrightarrow{G(f)} & FB \\ {\scriptstyle F(u_A)}\downarrow & & \downarrow{\scriptstyle F(u_B)} \\ FRA & \xrightarrow[F(f)]{} & FRB \end{array} ,
$$

and $F(u_A)$ and $F(u_B)$ are isos. Hence

*** $G(f)$ must equal $F(u_B)^{-1}F(f)F(u_A)$ if it exists.

Define G by $G = F$ on objects and G on maps is defined by ***. Then G is a functor and $GT = F$. Therefore, G exists when $\underline{C}$ is localizing.

Now we consider an abelian category $\underline{A}$ and a Serre subcategory $\underline{C}$. We want to construct a functor, T, and a category, $\underline{A}/\underline{C}$, so that $T: \underline{A} \longrightarrow \underline{A}/\underline{C}$ is exact, annihilates $\underline{C}$, and is universal for functors that are exact and annihilate $\underline{C}$.

<u>Theorem 2.1</u>. If $\underline{A}$ is well powered then $T: \underline{A} \longrightarrow \underline{A}/\underline{C}$ exists for any Serre subcategory $\underline{C}$.

<u>Proof</u>. (More or less due to Serre.)

Let objects of $\underline{A}/\underline{C}$ = objects of $\underline{A}$ and $TA = A$ on objects.

<u>Remark</u>. One way to obtain the morphism is to define $(A, B)_{\underline{A}/\underline{C}} = \varinjlim (A', B'')_{\underline{A}}$ where $\varinjlim$ is taken over

$$\begin{array}{ccccccc} 0 & \longrightarrow & A' & \longrightarrow & A & & \\ & & & \searrow & & & \\ & & B & \longrightarrow & B'' & \longrightarrow & 0 \end{array}$$

where the rows are exact and $\underline{C}$ isos. These form a direct system. This needs that the $\underline{C}$ subobjects and $\underline{C}$ quotient objects in $\underline{A}$ form a set so $\underline{C}$ small would take care of this. The difficulty in this method of proof lies in defining composition and proving associativity. We use a different method. Recall that any map $f: A \longrightarrow B$ can be looked at as $A \xrightarrow[(1,\ f)]{} A \times B \xrightarrow{pr_1} A$.

The image of $(1, f)$ is a subobject of $A \times B$ which is called the graph of f. This corresponds to finding a subobject Γ of A x B such that $\Gamma \to A \times B \to A$ is an iso. Hence to define $(A,B)_{\underline{A}/\underline{C}}$ we consider the following.

<u>Definition</u>. Let A and B $\in$ $\underline{A}$. Then a <u>pre C map</u> from A to B is a subobject Γ of A x B such that the composition $\Gamma \to A \times B \to A$ is a $\underline{C}$ iso.

<u>Remark</u>. The difficulty of this approach is that we could have Γ' a subobject of Γ such that $\Gamma' \to \Gamma$ is a $\underline{C}$ iso. They should give rise to the same map in $\underline{A}/\underline{C}$.

We introduce an equivalence relation on subobjects Γ of A x B which is generated by the relation "Γ' is a subobject of Γ and $\Gamma' \to \Gamma$ is a $\underline{C}$ iso."

<u>Definition</u>. $(A, B)_{\underline{A}/\underline{C}}$ is the set of equivalence classes. Note that we need $\underline{A}$ well powered to insure that $(A, B)_{\underline{A}/\underline{C}}$ is a set.

The next lemma connects our approach to the other one.

<u>Lemma 2.2</u>. (Goursat's Lemma) There exists a 1 - 1 correspondence between subobjects of A x B and maps of subobjects of A to quotient of B.

<u>Proof</u>. Let F be a subobject of A x B and let S = im $F \to A$ in the composition $F \to A \times B \to A$. Then we have $0 \to S \to A$ exact. We have the diagram

$$\begin{array}{ccccccccc} 0 & \to & K & \to & F & \to & S & \to & 0 \\ & & & & \downarrow{\scriptstyle pr_2} & & & & \\ & & & & B & & & & \\ & & & & \downarrow & & & & \\ & & & & \text{coker } f & & & & \end{array}$$

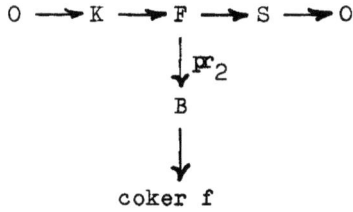

where f is the map $K \longrightarrow F \longrightarrow B$. This induces a map $S \longrightarrow \text{coker } f$. For modules $S = \{a \in A \mid \exists (a, b) \in F\}$, $K = \{(0, b) \in F\}$, $\text{im } f = \{b \in B \mid (0, b) \in F\}$, and $\text{coker } f = B/\text{im } f$.

To go the other way let $f: S \longrightarrow Q$ where S is a subobject of A and Q is a quotient of B. Form the pullback diagram

$$\begin{array}{ccc} P & \longrightarrow & S \times B \\ \downarrow & & \downarrow \\ S & \xrightarrow[(1,f)]{} & S \times Q \end{array} .$$

Then $P \longrightarrow S \times B \longrightarrow A \times B$ gives a subobject of $A \times B$. We check that these are two correspondences are inverses of each other by applying the embedding theorem and checking it in <u>Ab</u>.

Done with lemma 2.2.

<u>Definition</u>. A <u>relation between A and B</u> is a subobject of $A \times B$.

Let $R \subset A \times B$ and $S \subset B \times C$ be relations on modules. Then <u>$S \circ R$</u> $= \{(a, c) \mid \exists b$ such that $(a, b) \in R$ and $(b, c) \in S\}$.

This gives an associative composition. Let $R \subset A \times B$ and $S \subset B \times C$. Then form $A \times S$ and $R \times C$. For sets we could intersect them and remove B by projection and have $S \circ R$ be the image in $A \times C$ as indicated in the diagram

$$\begin{array}{ccc} (R \times C) \cap (A \times S) & \longrightarrow & A \times B \times C \\ & \searrow & \downarrow \\ \text{image is } S \circ R & & A \times C \end{array} .$$

But for sets intersection is pullback. In general we have

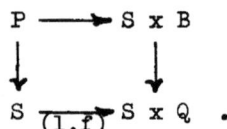

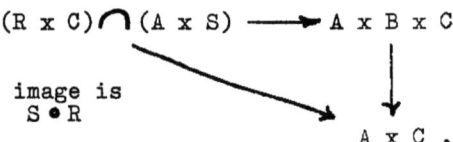

$$\begin{array}{ccccc} P & \longrightarrow & R \times C & & \\ \downarrow & & \downarrow & & \\ A \times S & \longrightarrow & A \times B \times C & & \\ & & & \searrow & \\ & & & & A \times C \end{array}$$

where P is the pullback and define image of P in A x C to be $S \circ R$. Then $T \circ (S \circ R) = (T \circ S) \circ R$ which we prove by applying the embedding theorem.

Using the isomorphism of A x B with B x A we define inverse of relation by

$$R^{-1} = \{(b, a) \mid (a, b) \in R\}.$$

Let R and R' be relations between A and B. Then $R \leq R'$ means $R \leq R'$ as subobjects.

Then we have $(S \circ R)^{-1} = R^{-1} \circ S^{-1}$ and

$R' \leq R$ implies $S \circ R' \leq S \circ R$,

$R' \circ T \leq R \circ T$, and

$R'^{-1} \leq R^{-1}$.

All of these follow easily from the embedding theorem.

Now let R be a relation between A and B and X be a subobject of A. We want to define $R(\underline{X})$. For sets $R \longrightarrow A \times B$ and $X \subset A$ the image of $(X \times B) \cap R \longrightarrow B$ is the image, R(X). So in general we form the pullback P and take image in B

$$\begin{array}{ccccc} P & \longrightarrow & X \times B & & \\ \downarrow & & \downarrow & & \\ R & \longrightarrow & A \times B & & \\ & & & \searrow & \\ & & & & B\ . \end{array}$$

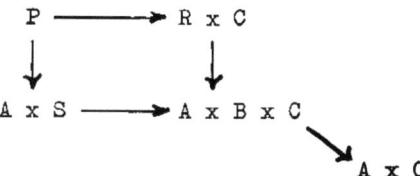

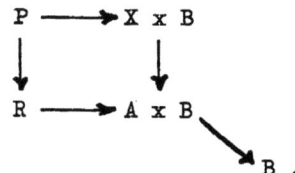

If $R \subset A \times B$, $S \subset B \times C$, and $X \subset A$, then $S(R(X)) = (S \circ R)(X)$. We prove this by applying the embedding theorem and doing it for sets.

<u>Definition</u>. If R is a relation between A and B then the <u>domain of R</u> is $R^{-1}(B)$, and the <u>codomain</u> of R is R(A).

$R^{-1} \circ R \geqslant \Delta_D$ where D is the domain of R and Δ_D is the image of D in A x A by the diagonal map. This is checked by the embedding theorem.

Now we want a way of distinguishing the pre $\underline{A}/\underline{C}$ maps from arbitrary subobjects of A x B.

<u>Lemma 2.3</u>. Let R be a relation between A and B. Then R is a pre $\underline{A}/\underline{C}$ map if and only if

1) $X \subset A$ and $X \in \underline{C}$ implies $R(X) \in \underline{C}$ and

2) $Y \subset B$ and $B/Y \in \underline{C}$ implies $A/R^{-1}(Y) \in \underline{C}$.

<u>Proof</u>. $\Longrightarrow$ 1) Let $R \subset A \times B$ be a pre $\underline{A}/\underline{C}$ map. Then $R \longrightarrow A$ is a $\underline{C}$ iso. Let X be a subobject of A with $X \in \underline{C}$. Form the pullback diagram

$$\begin{array}{ccccc} K & & K & & \\ \downarrow & & \downarrow & & \\ P & \longrightarrow & R & \longrightarrow & B \\ \downarrow & & \downarrow & & \\ X & \longrightarrow & A & & \end{array}$$

where K is the ker of $R \longrightarrow A$.

$K \in \underline{C}$ since $R \longrightarrow A$ is a $\underline{C}$ iso. $\mathrm{Im}(P \longrightarrow X) \in \underline{C}$ since $X \in \underline{C}$ and $\underline{C}$ is closed under subobjects. Thus $P \in \underline{C}$. Hence, all quotients of P are in $\underline{C}$ and $R(X) \in \underline{C}$.

2) Let Y be a subobject of B with $B/Y \in \underline{C}$. Forming the pullback we obtain

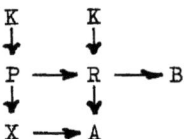

$$\begin{array}{ccccccccc} 0 & \to & P & \to & R & \to & B/Y & \to & 0 \\ & & \downarrow & & \downarrow & & \downarrow & & \\ 0 & \to & Y & \to & B & \to & B/Y & \to & 0 \end{array} .$$

$B/Y \in \underline{C}$ implies $P \to R$ is a $\underline{C}$ iso. Hence $P \to R \to A$ is a $\underline{C}$ iso. Thus coker $(P \to A) \in \underline{C}$. But coker $(P \to A) = A/R^{-1}(Y)$.

$\Leftarrow$ We want that 1) and 2) implies $R \to A$ is a $\underline{C}$ iso. $R(0) \in \underline{C}$ by 1) and $A/R^{-1}(B) \in \underline{C}$ by 2). Form the pullback

$$\begin{array}{ccccc} 0 & \to & P & \to & R \\ & & \downarrow & & \downarrow \\ 0 & \to & 0 & \to & A \end{array} .$$

Then $P = \ker (R \to A)$. Hence, $P \to R \to B$ is a mono. So $R(0) = \mathrm{im}[P \to B]$ is isomorphic to P and $R(0) \in \underline{C}$. Thus $R \to A$ is a $\underline{C}$ mono.

$$\begin{array}{ccccc} P & \xrightarrow{=} & R & \to & A \\ \downarrow & & \downarrow & & \\ B & \to & B & & \end{array}$$

shows that $R^{-1}(B) = \mathrm{im}(P \to A)$. $A/R^{-1}(B) = \mathrm{coker}\ (P \to A) \in \underline{C}$ by 2). Thus $R \to A$ is a $\underline{C}$ epi.

Done with lemma 2.3.

Corollary 2.4. If R and S are pre $\underline{C}$ maps then so is $S \circ R$.

Proof. Clear from lemma 2.3.

Now we check that composition is compatible with identification.

Lemma 2.5. If R is equivalent to R' and S is equivalent to S', then $S \circ R$ is equivalent to $S' \circ R'$ when $S \circ R$ is defined.

Proof. The equivalence relation is generated by $\leqslant$ so we only need to check that $R \leqslant R'$ implies $S \circ R$ is equivalent to $S \circ R'$

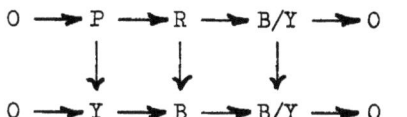

and $S \leq S'$ implies $S \circ R$ is equivalent to $S' \circ R$ and finish by induction on number of steps.

Thus we have a well defined associative composition

$$(A, B)_{\underline{A}/\underline{C}} \times (B, C)_{\underline{A}/\underline{C}} \longrightarrow (A, C)_{\underline{A}/\underline{C}} .$$

$\Delta_A = \Gamma_{1_A} = I \subset A \times A$ clearly gives rise to the class of a two sided identity. Thus, $\underline{A}/\underline{C}$ is a category.

We define $T: \underline{A} \longrightarrow \underline{A}/\underline{C}$ by $T(A) = A$ on objects. Let $f \in (A, B)_{\underline{A}}$. Then f yields its graph $\Gamma_f \subset A \times B$. Γ_f is a pre $\underline{A}/\underline{C}$ map. Define $T(f) = [\Gamma_f]$. This clearly makes T into a covariant functor. We finally need to check that $T: \underline{A} \longrightarrow \underline{A}/\underline{C}$ has the desired properties.

Claim. If $F: \underline{A} \longrightarrow \underline{B}$ is an exact functor and F annihilates $\underline{C}$ then there exists a functor G such that $F = GT$.

Define $G(A) = F(A)$. Then $F = GT$ on objects. Let $R \longrightarrow A \times B$ be a pre $\underline{A}/\underline{C}$ map. Then $FR \longrightarrow FA \times FB$ and $R \longrightarrow A$ is a $\underline{C}$ iso. Then $FR \longrightarrow FA$ is an iso since F annihilates $\underline{C}$. Hence, $FR = \Gamma_g$ for a unique $g: FA \longrightarrow FB$. Define G on maps by $G[R] = g$ where $\Gamma_g = F(R)$.

We have to check that this is compatible with identification. As before it is enough to show that if $R' \leq R \longrightarrow A \times B$ then they give rise to the same $g: FA \longrightarrow FB$. We have the diagram

$$\begin{array}{ccc} R & \longrightarrow & A \\ \uparrow & \nearrow & \\ R' & & \end{array}$$

with both maps to A being $\underline{C}$ isos.

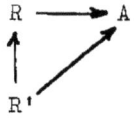

Applying F yields

$$\begin{array}{ccccc} FR & \longrightarrow & FA \times FB & \longrightarrow & FA \\ \uparrow & & & \nearrow & \\ FR' & & & & \end{array}$$

such that FR $\longrightarrow$ FA and FR' $\longrightarrow$ FA are both isos. Hence FR' $\longrightarrow$ FR is an iso. Therefore, they give rise to the same subobject of FA x FB and, hence, to the same map g: FA $\longrightarrow$ FB. So G[R] = G[R'] and G is well defined on maps. GT(f) = F(f) is clear.

Let R $\longrightarrow$ A x B and S $\longrightarrow$ B x C. We defined composition by taking pullback and then image in the diagram

$$\begin{array}{ccccc} P & \longrightarrow & A \times S & & \\ \downarrow & & \downarrow & & \\ R \times C & \longrightarrow & A \times B \times C & & \\ & & & \searrow & \\ & & & & A \times C. \end{array}$$

Applying F yields

$$\begin{array}{ccccc} FP & \longrightarrow & F(A \times S) & & \\ \downarrow & & \downarrow & & \\ F(R \times C) & \longrightarrow & F(A \times B \times C) & & \\ & & & \searrow & \\ & & & & F(A \times C). \end{array}$$

Hence composition is preserved and G is a functor with the claimed properties.

Proposition 2.6. Let G and G' be functors from A/C to B where B is any category. If GT = G'T then G = G'.

Remark. This says that T acts like an epimorphism of categories.

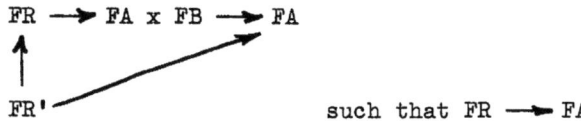

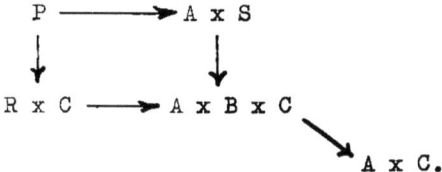

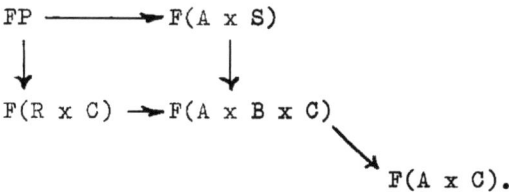

<u>Proposition 2.7</u>. Let G and G' be functors from <u>A</u>/<u>C</u> to <u>B</u> with GT naturally isomorphic to G'T, then G is naturally isomorphic to G'.

<u>Lemma 2.8</u>. If f: A $\longrightarrow$ B in <u>A</u> and f is a <u>C</u> iso then T(f) is an iso in <u>A</u>/<u>C</u>.

<u>Proof</u>. Γ_f, the graph of f, is a pre <u>A</u>/<u>C</u> map. We have the diagram

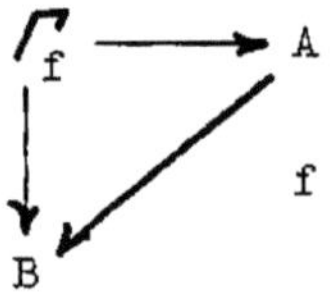

where $\Gamma_f \longrightarrow A$ is an iso and $\Gamma_f \longrightarrow B$ is a <u>C</u> iso. Hence, $(\Gamma_f)^{-1} \longrightarrow A \times A$ is a pre <u>A</u>/<u>C</u> map and, therefore, represents $h \in (B, A)_{\underline{A}/\underline{C}}$. h is the required inverse since hT(f) is represented by $\Gamma_f^{-1}\Gamma_f$ which is in the class of the identity. Similarly T(f)h is the identity. Done with lemma 2.8.

<u>Lemma 2.9</u>. Let h: TA $\longrightarrow$ TB in <u>A</u>/<u>C</u> then there exists a diagram

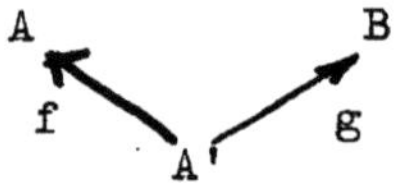

in <u>A</u> with f a <u>C</u> iso such that

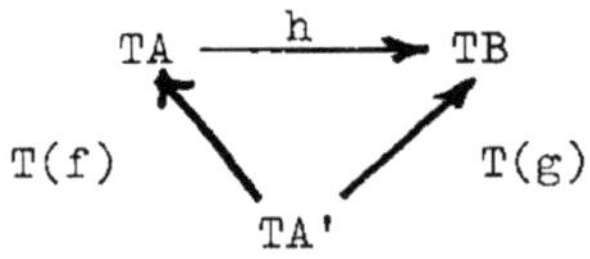

commutes.

<u>Proof</u>. Pick R representing h. Then R $\longrightarrow$ A x B and R $\longrightarrow$ A is a <u>C</u> iso. Let A' = R and f and g be the projection onto A and B

respectively. Then f is a $\underline{C}$ iso. We claim $hT(f) = T(g)$. That is $R \circ \Gamma_f \geq \Gamma_g$. But we can even prove that if R is a relation between A and B and we defined f and g as above then $R \circ \Gamma_f \geq \Gamma_g$ by applying the embedding theorem. Done with lemma 2.9.

Now we return to the proofs of propositions 2.6 and 2.7.

Let G and G' be two functors from $\underline{A}/\underline{C}$ to $\underline{B}$ such that $GT = G'T$. Then $G = G'$ on objects. Let $h: TA \longrightarrow TB$. Apply lemma 2.9 and produce f, g, and A'

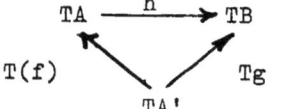

$$\begin{array}{ccc} A & & B \\ & {}^{f}\nwarrow \quad \nearrow^{g} & \\ & A' & \end{array}$$

where f is a $\underline{C}$ iso.

Apply T yielding

$$* \qquad \begin{array}{ccc} TA & \xrightarrow{h} & TB \\ & {}^{T(f)}\nwarrow \quad \nearrow^{Tg} & \\ & TA' & \end{array}$$

where T(f) is an iso.

Apply G

$$\begin{array}{ccc} GT(A) & \xrightarrow{G(h)} & GT(B) \\ & {}^{GT(f)}\nwarrow \quad \nearrow^{GT(g)} & \\ & GT(A') & \end{array}$$

Apply G' to * gives

$$\begin{array}{ccc} G'T(A) & \xrightarrow{G'(h)} & G'T(B) \\ & {}^{G'T(f)}\nwarrow \quad \nearrow^{G'T(g)} & \\ & G'T(A') & \end{array} .$$

But $G'T = GT$ and $GT(f) = G'T(f)$ is an iso. Therefore $G(h) = G'(h)$ and proposition 2.6 is done.

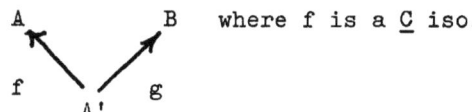

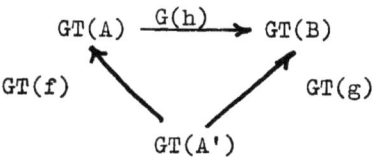

Let G' and G: $\underline{A}/\underline{C} \longrightarrow \underline{B}$ with G'T naturally isomorphic to GT. Then we have

$$\eta_A : GT(A) \xrightarrow{\cong} G'T(A) \text{ for all } A \in \underline{A}.$$

Let f: $A \longrightarrow B$. We get the diagram

$$\begin{array}{ccc} GT(A) & \xrightarrow{\cong} & G'T(A) \\ {\scriptstyle GT(f)}\downarrow & & \downarrow{\scriptstyle G'T(f)} \\ GT(B) & \overset{\cong}{\text{———}} & G'T(B) \end{array} \quad .$$

We need $h_A : G(A) \xrightarrow{\cong} G'(A)$.

Let $h \in (A, B)_{\underline{A}/\underline{C}}$. Then we have

$$\begin{array}{ccc} G(A) & \xrightarrow{\cong} & G'(A) \\ {\scriptstyle G(h)}\downarrow & & \downarrow{\scriptstyle G'(h)} \\ G(B) & \overset{\cong}{\text{———}} & G'(B) \end{array} \quad .$$

We have $\eta_A : G(A) \xrightarrow{\cong} G'(A)$ since TA = A. These work for h_A. Hence G' is naturally isomorphic to G.

Done with proposition 2.7.

Now we begin proving that $\underline{A}/\underline{C}$ is an abelian category. First we show that $\underline{A}/\underline{C}$ is additive and that T is additive. By examining the case of modules we arrive at the following definition.

Definition. Let R' and R" be pre $\underline{A}/\underline{C}$ maps from A to B. Then $\underline{R' + R''} = \{(a, b) \mid \exists b', b'' \in B \text{ with } b = b' + b'', (a, b') \in R', \text{ and } (a, b'') \in R''\}$.

We examine a pullback-image diagram as before.

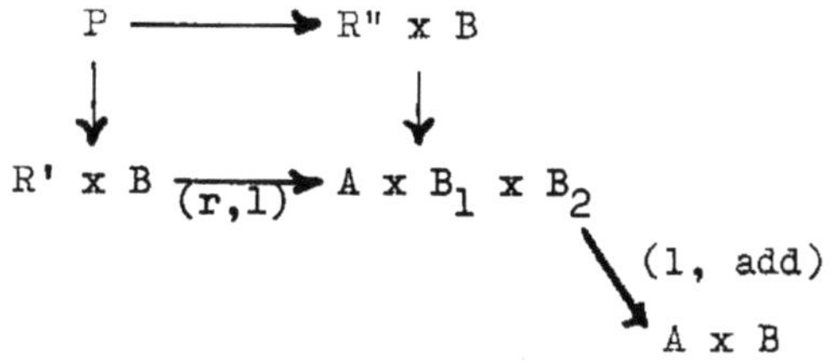

where add is the addition map and the map $R'' \times B \longrightarrow A \times B_1 \times B_2$ sends R'' to A and B_2 by projection and B to B_1 by the identity.

Im $P \longrightarrow A \times B$ is $R' + R''$ as defined above. This addition is associative, has a unit, and has inverses. $R' \leq R''$ implies $R + R' \leq R + R''$. All of this is checked by applying the embedding theorem. Addition is well defined follows as before. Thus $\underline{A}/\underline{C}$ is additive and T is additive except we need to check that if R and R' are pre $\underline{C}$ maps then so is $R + R'$. We apply the criterion of lemma 2.3. We claim

*) $(R' + R'')(X) \leq R'(X) + R''(X)$

**) $(R' + R'')^{-1}(Y) \geq R'^{-1}(Y) \cap R''^{-1}(Y)$.

Both of these are preserved by embedding so we embed.

$$(R' + R'')(X) = \{b \mid \exists (a, b) \in R' + R'' \text{ with } a \in X\}$$
$$= \{b \mid \exists (a, b') \in R', (a, b'') \in R'', b' + b'' = b, \text{ and } a \in X\}.$$
$$R'(X) + R''(X) = \{b \mid b = b' + b'', b' \in R'(X), b'' \in R''(X)\}$$
$$= \{b \mid \exists (a', b') \in R', (a'', b'') \in R'', a', a'' \in X, \text{ and } b = b' + b''\}.$$

Hence *) is proved.

$$(R' + R'')^{-1}(Y) = \{a \mid \exists (b, a) \in (R' + R'')^{-1} \text{ and } b \in Y\}$$
$$= \{a \mid \exists (a, b') \in R', (a, b'') \in R'', \text{ and } b' + b'' = b \in Y\}.$$
$$R'^{-1}(Y) \cap R''^{-1}(Y) = \{a \mid \exists (b', a) \in R'^{-1}, b' \in Y \text{ and } \exists (b'', a) \in R''^{-1}, b'' \in Y\}.$$

But b' and $b'' \in Y$ implies $b' + b'' \in Y$ so **) follows.

Now to apply lemma 2.3. 1) Let $X \subset A$ with $X \in \underline{C}$. Then $(R' + R'')(X) \leq R'(X) + R''(X)$. But $R'(X) \in \underline{C}$ and $R''(X) \in \underline{C}$. Hence $R'(X) \oplus R''(X) \in \underline{C}$. Thus, $R'(X) + R''(X) \in \underline{C}$ since $\underline{C}$ has images. Finally $(R' + R'')(X) \in \underline{C}$ since $\underline{C}$ has subobjects.
2) Let $Y \subset B$ with $B/Y \in \underline{C}$. Then $A/R'^{-1}(Y) \in \underline{C}$ and $A/R''^{-1}(Y) \in \underline{C}$ implies $A/R'^{-1}(Y) \oplus A/R''^{-1}(Y) \in \underline{C}$. $A \twoheadrightarrow A/R'^{-1}(Y) \oplus A/R''^{-1}(Y)$ has kernel $R'^{-1}(Y) \cap R''^{-1}(Y)$. $\underline{C}$ closed under subobjects means the image of $A = A/R'^{-1}(Y) \cap R''^{-1}(Y) \in \underline{C}$. But $(R' + R'')^{-1}(Y) \geq R'^{-1}(Y) \cap R''^{-1}(Y)$ means there is an epi

$A/R'^{-1}(Y) \cap R''^{-1}(Y) \twoheadrightarrow A/(R' + R'')^{-1}(Y)$ and $\underline{C}$ closed under quotients implies $A/(R' + R'')^{-1}(Y) \in \underline{C}$. Therefore, addition of pre $\underline{A/C}$ maps is a pre $\underline{A/C}$ map.

Now we define addition in $(A, B)_{\underline{A/C}}$. Let $f', f'' \in (A, B)_{\underline{A/C}}$. Pick representatives R', R''. Then $R' + R''$ represents $f \in (A, B)_{\underline{A/C}}$.

Then $f = f' + f''$ is well defined. The negative of a pre $\underline{A/C}$ map is a pre $\underline{A/C}$ map. Composition is bilinear. All of these are checked by embedding. The last follows from

$$(R' + R'') \circ S \leq R' \circ S + R'' \circ S \quad \text{and}$$
$$T \circ (R' + R'') \geq T \circ R' + T \circ R'' .$$

Clearly T: $\underline{A} \to \underline{A}/\underline{C}$ is additive.

Next we check that $\underline{A}/\underline{C}$ has zero objects, finite sums, and finite products.

Let 0 be a zero object in $\underline{A}$. Then 0 is a zero object in $\underline{A}/\underline{C}$.

$1_0 = 0$ in $\underline{A}$ is given. Then

$1_{T(0)} = T(1_0) = T(0) = 0_{T(0)}$ since T is an additive functor. Hence T(0) is a zero object in $\underline{A}/\underline{C}$.

Let A, B $\in \underline{A}$. We can form direct sum diagrams in $\underline{A}$ which T preserves since T is additive. Thus $\underline{A}/\underline{C}$ has finite sums and products.

<u>Lemma 2.10</u>. Let f: $A \to B$ be in $\underline{A}$. Then $T(f) = 0$ if and only if im f $\in \underline{C}$.

<u>Proof</u>. $\Leftarrow$ We factor f as

$A \to \text{im } f \to B$. Applying T gives

$TA \to T(\text{im } f) \to TB$. If $T(\text{im } f) = 0$

then T(f) factors through 0, and, hence, $T(f) = 0$.

So we need to show $C \in \underline{C}$ implies $T(C) = 0$. The inclusion $0 \to C \times C$ is a pre $\underline{A}/\underline{C}$ map. $0 \to C$ is a $\underline{C}$ iso. Hence every pre $\underline{A}/\underline{C}$ map from C to C is equivalent to the zero map. In particular $1_{r_c} = 0_{r_c}$ or C is a zero object.

$\Rightarrow$ To prove this we need another lemma.

<u>Lemma 2.11</u>. Let R' and R" $\subset A \times B$ be pre $\underline{A}/\underline{C}$ maps. Then R is equivalent to R' if and only if there exists a pre $\underline{A}/\underline{C}$ map R with $R' \geqslant R \leqslant R''$.

<u>Remark</u>. By applying lemma 2.11 to any representative for f and the representative $A \times 0$ we will finish lemma 2.10.

<u>Proof</u>. $\Leftarrow$ by definition.

$\Rightarrow$ Consider the relation on pre $\underline{A}/\underline{C}$ maps given by $R' \sim R''$ if and only if there exists a pre $\underline{A}/\underline{C}$ map R with $R' \geq R \leq R''$. If this is an equivalence relation we are done. It is clearly symmetric and reflexive. Suppose we are given R', R, R'', S and S'' all pre $\underline{A}/\underline{C}$ maps with

$$R' \geq R \leq R'' \geq S \leq S''.$$

We want to collapse the middle. $R \leq R'' \geq S$ to get $R \geq R''' \leq S$. That is we want to show that if $R \leq T \geq S$ all pre $\underline{A}/\underline{C}$ maps then there exists a pre $\underline{A}/\underline{C}$ map, U, with $R \geq U \leq S$. That is we want to fill in the diagram

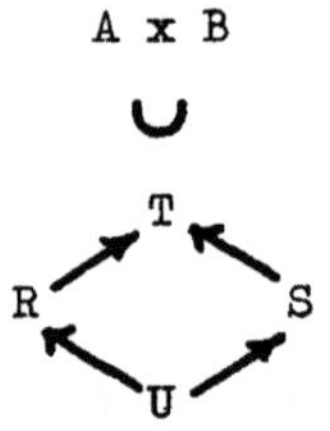

This suggests we try the pullback (which equals the intersection) for it. It is easy to see that it works.

Done with lemmas 2.10 and 2.11.

Now we want to show that $\underline{A}/\underline{C}$ has kernels and cokernels, T preserves kernels and cokernels, and that coim is isomorphic to image in $\underline{A}/\underline{C}$. If we can show that our situation is self dual, then we will only need to produce kernels and show that T preserves them. Hence, we need the dual of lemma 2.9. That is given $h: TA \longrightarrow TB$ we need that there exists

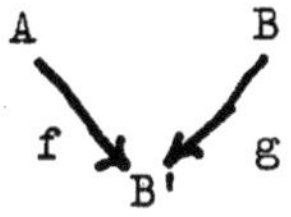

where g is a $\underline{C}$ iso.

Apply lemma 2.9 we obtain

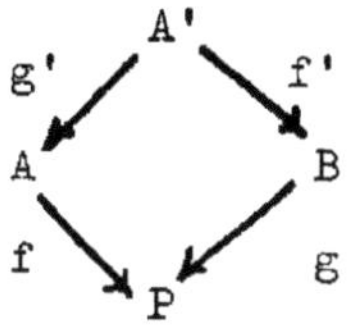

where g' is a $\underline{C}$ iso and P is the pushout. But, by the pushout theorem, g' a $\underline{C}$ iso implies g is a $\underline{C}$ iso. Thus our situation is self dual.

To investigate the existence and preservation of kernels we first investigate a slightly weaker concept.

<u>Definition</u>. Let f be a morphism from A to B in any abelian category. Then i: K $\longrightarrow$ A is a <u>pseudo kernel</u> if for any map j: D $\longrightarrow$ A with fj = 0 there exists a map g: D $\longrightarrow$ K such that the following diagram commutes:

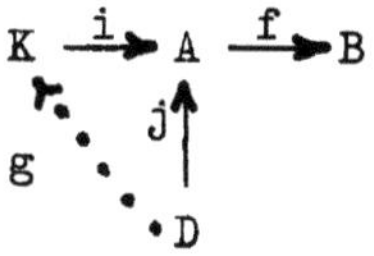

.

This differs from the definition of kernel in that we do not assume that g is unique.

Next we show T preserves pseudo kernels. Let i: K $\longrightarrow$ A be a pseudo kernel for f. Consider the diagram

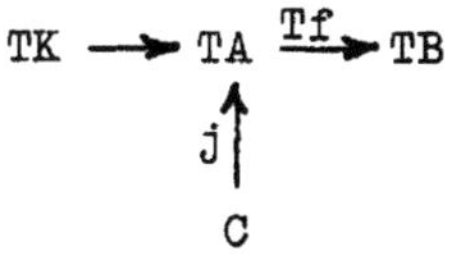

where $(Tf)j = 0$. C must be T of something. Say it is $T(D)$. So we have the diagram

$$\begin{array}{ccccc} TK & \longrightarrow & TA & \xrightarrow{Tf} & TB \\ & & {\scriptstyle j}\uparrow & & \\ & & TD & & \end{array} .$$

Applying lemma 2.9 we obtain the diagram

$$\begin{array}{ccc} D & & A \\ {\scriptstyle \chi}\nwarrow & & \nearrow{\scriptstyle m} \\ & D' & \end{array} \qquad \text{where } \chi \text{ is a } \underline{C} \text{ iso}$$

and with

$$\begin{array}{ccc} TD & \xrightarrow{j} & TA \\ {\scriptstyle T(\chi)}\nwarrow & & \nearrow{\scriptstyle T(m)} \\ & TD' & \end{array} .$$

That is we can replace D by D' and assume $j = T(\chi)$. Then $T(fj) = 0$ and, hence, im $fj \in \underline{C}$. We form the sequence in $\underline{A}$.

$0 \longrightarrow \ker fj \xrightarrow{i'} D \longrightarrow \operatorname{im} fj \longrightarrow 0$. i' is a $\underline{C}$ iso. Let $\ker fj = D'$.

Now we have the commutative diagram

$$\begin{array}{ccccc} K & \xrightarrow{i} & A & \xrightarrow{f} & B \\ & {\scriptstyle \chi'}\nearrow & \uparrow{\scriptstyle \chi} & & \\ D' & \xrightarrow[i']{} & D & & \end{array} \qquad \text{where } f\chi' = 0.$$

Applying T we obtain

$$\begin{array}{ccccc} TK & \xrightarrow{Ti} & TA & \xrightarrow{Tf} & TB \\ & {\scriptstyle T\chi'}\uparrow & & \nwarrow{\scriptstyle j} & \\ & TD' & \xrightarrow[Ti']{} & TD. & \end{array}$$

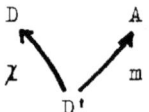

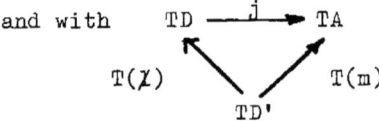

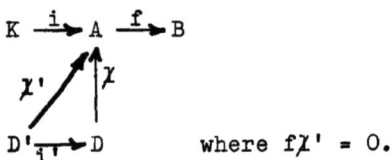

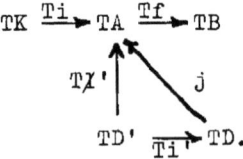

By replacing D by D' we can assume $f\chi = 0$. Then there exists $k: D \to K$ such that $ik = \chi$. Hence $j = T'\chi$ factors through TK. Therefore, T preserves pseudo kernels.

Let $0 \to A \xrightarrow{f} B$ be exact. Then T0 is a pseudo kernel of Tf. But $T0 = 0$. Therefore Tf is mono and T preserves monos.

<u>Remark</u>. Let $f: A \to B$ in $\underline{A}$. Then $K \xrightarrow{i} A$ is a kernel for f if and only if it is a pseudo kernel and a mono. We obtain liftings since it is a kernel and they are unique since it is a mono.

Let $h: TA \to TB$ be a map in $\underline{A}/\underline{C}$. By lemma 2.9 we obtain

$$\begin{array}{ccc} A & & B \\ {\scriptstyle f}\nwarrow & & \nearrow{\scriptstyle g} \\ & A' & \end{array}$$

where f is a $\underline{C}$ iso and the following diagram commutes:

$$\begin{array}{ccc} TA & \xrightarrow{h} & TB \\ {\scriptstyle Tf}\nwarrow & & \nearrow{\scriptstyle Tg} \\ & TA' & \end{array} .$$

Since Tf is an iso, h has a kernel if Tg does. But T preserves kernels since it preserves monos and pseudo kernels. Therefore T(ker g) is a kernel for Tg. Hence h has a kernel and T preserves kernels. By duality $\underline{A}/\underline{C}$ has cokernels and T preserves them.

Let $h: TA \to TB$ in $\underline{A}/\underline{C}$. We want coim h is isomorphic to im h. We get the usual diagram

$$\begin{array}{ccc} TA & \xrightarrow{h} & TB \\ {\scriptstyle Tf}\nwarrow & & \nearrow{\scriptstyle Tg} \\ & TA' & \end{array}$$

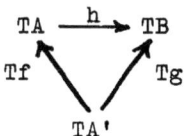

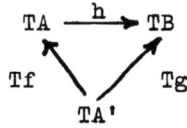

where Tf is an iso. Hence it is enough to show that coim Tg is isomorphic to image g. But coim g is isomorphic to im g and T preserves coim and im. Thus coim and im are isomorphic in $\underline{A}/\underline{C}$. Hence, $\underline{A}/\underline{C}$ is an abelian category and T is exact.

Let G: $\underline{A}/\underline{C} \longrightarrow B$ be a functor with B an abelian category. Suppose GT is exact. Let h: $TA \longrightarrow TB$ in $\underline{A}/\underline{C}$. Obtain

$$\begin{array}{ccc} TA & \xrightarrow{h} & TB \\ {\scriptstyle Tf}\nwarrow & & \nearrow{\scriptstyle Tg} \\ & TA' & \end{array}$$

where Tf is an iso.

Apply G

$$\begin{array}{ccc} GTA & \xrightarrow{G(h)} & GTB \\ {\scriptstyle GT(f)}\nwarrow & & \nearrow{\scriptstyle GT(g)} \\ & GTA' & \end{array} .$$

Hence, to check if G preserves ker and coker we can assume $h = T(g)$.

Then $\ker h = T(\ker g)$ and

$$G(\ker h) = GT(\ker g) = \ker(GT(g)) = \ker G(h)$$

similarly for coker.

Thus if GT if $\begin{matrix}\text{right}\\ \text{left}\end{matrix}$ exact so is G. Done with theorem 2.1.

<u>Corollary 2.12</u>. If $0 \longrightarrow TA' \longrightarrow TA \longrightarrow TA'' \longrightarrow 0$ is an exact sequence in $\underline{A}/\underline{C}$ then there is an exact sequence $0 \longrightarrow B' \longrightarrow B \longrightarrow B'' \longrightarrow 0$, in $\underline{A}$ such that the following diagram commutes with all vertical maps isos:

$$\begin{array}{ccccccccc} 0 & \longrightarrow & TB' & \longrightarrow & TB & \longrightarrow & TB'' & \longrightarrow & 0 \\ & & \downarrow & & \downarrow & & \downarrow & & \\ 0 & \longrightarrow & TA' & \longrightarrow & TA & \longrightarrow & TA'' & \longrightarrow & 0 \end{array}$$

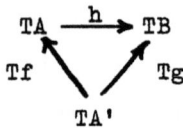

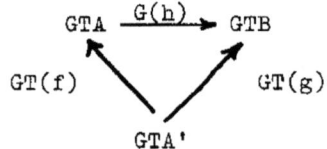

Proof. By lemma 2.9 there is a diagram

$$\begin{array}{ccc} & B & \\ \swarrow & & \searrow \\ A & & A'' \end{array}$$

such that

$$\begin{array}{ccc} & TB & \\ {\scriptstyle\approx}\swarrow & & \searrow \\ TA & \longrightarrow & TA'' \end{array}$$

commutes.

Let B" be the image of $B \longrightarrow A''$ and B' the kernel of $B \longrightarrow B''$. Since T is exact we see easily that $TB'' \xrightarrow{\approx} TA''$.

Theorem 2.13. Let $\underline{A}$ be an abelian category, $\underline{C}$ a Serre subcategory such that $\underline{A}/\underline{C}$ exists. Then $\underline{C}$ is localizing if and only if $T: \underline{A} \longrightarrow \underline{A}/\underline{C}$ has a right adjoint.

Proof. $\Longrightarrow$ Let $\underline{L}$ be the full subcategory of $\underline{C}$ closed objects and $R: \underline{A} \longrightarrow \underline{L}$ be the reflection. We have the diagram

$$\begin{array}{ccc} \underline{A} & \xrightarrow{R} & \underline{L} \\ {\scriptstyle T}\searrow & & \nearrow{\scriptstyle E} \\ & \underline{A}/\underline{C} & \end{array}$$

where E is an equivalence and T is naturally isomorphic to ER. R is the left adjoint of $i: \underline{L} \longrightarrow \underline{A}$; and, hence, T has a right adjoint.

$\Longleftarrow$ Let S and T be a pair of adjoint functor $\underline{A} \underset{S}{\overset{T}{\rightleftarrows}} \underline{A}/\underline{C}$. Then we have

$$(A, SB) = (TA, B) \text{ where } A \in \underline{A} \text{ and } B \in \underline{A}/\underline{C}.$$

First we show that for all $B \in \underline{A}/\underline{C}$ SB is $\underline{C}$ closed. Let $u: A \longrightarrow A'$ be a $\underline{C}$ iso. Then we want $(A, SB) \xleftarrow{(u, 1)} (A', SB)$ to be an iso.

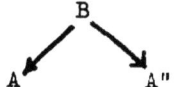

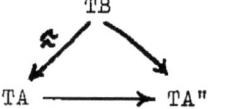

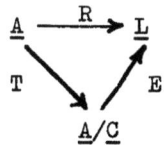

By adjointness this is the same as $(TA, B) \xleftarrow[(TU,1)]{} (TA', B)$. But TU is an iso. Hence, the map on Hom is, and SB is $\underline{C}$ closed.

Let $A \in \underline{A}$. Then $(TA, TA) = (A, STA)$. Let η_A correspond to 1_{TA}. Then $\eta_A : A \longrightarrow ST(A)$ and $ST(A)$ is $\underline{C}$ closed. To show $\underline{C}$ is localizing, it is enough to show that η_A is a $\underline{C}$ mono by theorem 1.7. Let $B \in \underline{A}/\underline{C}$. Then $(SB, SB) = (TSB, B)$ and ν_B corresponds to 1_{SB} where $\nu_B : TSB \longrightarrow B$. Then the composition

$$TA \xrightarrow[T(\eta_A)]{} TSTA \xrightarrow[\nu_{TA}]{} TA$$

is the identity. Therefore, $T(\eta_A)$ is a mono. Let K be the kernel of $A \longrightarrow ST(A)$. Then $K \in \underline{C}$ since $T(A) \longrightarrow TST(A)$ is a mono. Hence η_A is a $\underline{C}$ mono. Done.

PART II

K-THEORY

Chapter 1. Definition of $K_0(\underline{A})$ and some examples.

Definition. Let $\underline{A}$ be a small abelian category. $K_0(\underline{A})$ is the abelian group generated by $[A]$, where A runs over the objects of $\underline{A}$ and relation $[A] = [A'] + [A'']$ for each short exact sequence $0 \longrightarrow A' \longrightarrow A \longrightarrow A'' \longrightarrow 0$ in $\underline{A}$.

Let G be any abelian group and $f\colon \mathrm{obj}(\underline{A}) \longrightarrow G$ such that for each short exact sequence $0 \longrightarrow A' \longrightarrow A \longrightarrow A'' \longrightarrow 0$ in $\underline{A}$ we have $f(A) = f(A') + f(A'')$. Then this factors through the map to $K_0(\underline{A})$ uniquely. That is $K_0(\underline{A})$ is the universal function of the above type.

Remark. Investigation of $K_0(\underline{A})$ stems from the original example of Grothendieck given by the construction: X is an algebraic variety, $\underline{A}$ is the category of coherent sheaves, and $f(A)$ is $\chi(X, A)$, the Euler characteristic of the sheaf A.

Proposition 1.1. The following are immediate consequences of the definition of $K_0(\underline{A})$:

1) $[0] = 0$,
2) If A is isomorphic to B, then $[A] = [B]$, and
3) $[A \oplus B] = [A] + [B]$.

Proof.

1) $0 \longrightarrow 0 \longrightarrow 0 \longrightarrow 0 \longrightarrow 0$ is exact.
2) $0 \longrightarrow 0 \longrightarrow A \longrightarrow B \longrightarrow 0$ is exact.
3) $0 \longrightarrow A \longrightarrow A \oplus B \longrightarrow B \longrightarrow 0$ is exact. Done.

Proposition 1.2. Let k be a field and $\underline{A}$ be the category of finite dimensional vector spaces over k, then $K_0(\underline{A}) = Z$.

Proof. Let V be a vector space of dimension n. Then $V \cong k^n$ implies $[V] = n[k]$. Hence $[k]$ generates $K_0(\underline{A})$. Consider the function dim: object $\underline{A} \longrightarrow Z$. If $0 \longrightarrow V' \longrightarrow V \longrightarrow V'' \longrightarrow 0$ is an exact sequence of vector spaces, then $\dim V = \dim V' + \dim V''$. Hence, there is a homomorphism from $K_0(\underline{A})$ onto Z. Therefore, $K_0(\underline{A}) = Z$.

Done.

Proposition 1.3. If $\underline{A}$ has countable direct sums (or products), then $K_0(\underline{A}) = 0$.

Proof. Let $A \in \underline{A}$ and $B = \coprod_1^\infty A$. Then $A \oplus B$ is isomorphic to B. Thus $[B] = [A \oplus B] = [A] + [B]$. Hence $[A] = 0$ and $K_0(\underline{A}) = 0$. (If $\underline{A}$ has countable products, set $B = \prod_1^\infty A$). Done.

If $\underline{A}$ and $\underline{B}$ are small abelian categories and T is an exact functor from $\underline{A}$ to $\underline{B}$, then the map $[A] \rightsquigarrow [T(A)]$ induces a homomorphism $K_0(T)\colon K_0(\underline{A}) \longrightarrow K_0(\underline{B})$. Hence K_0 is a covariant functor from the category of small abelian categories and exact functors to $\underline{Ab}$.

If S and T are naturally isomorphic functors, then $K_0(S) = K_0(T)$. For, SA isomorphic to TA implies $K_0(S)([A]) = [SA] = [TA] = K_0(T)([A])$.

Proposition 1.4. If $\underline{A}$ and $\underline{B}$ are equivalent small abelian categories then $K_0(\underline{A})$ is isomorphic to $K_0(\underline{B})$.

Proof. Let S and T be the equivalences. $\underline{A} \underset{T}{\overset{S}{\rightleftarrows}} B$. Then ST is naturally isomorphic to $id_{\underline{B}}$ and TS is naturally isomorphic to $id_{\underline{A}}$. Therefore, $K_0(S)K_0(T) = 1_{K_0(\underline{B})}$ and $K_0(T)K_0(S) = 1_{K_0(\underline{A})}$.

Done.

Definition. If an abelian category $\underline{A}$ is equivalent to a small abelian category $\underline{B}$, $K_0(\underline{A})$ is $K_0(B)$.

Remark. This is well defined; and if $\underline{A}$ is already small, the two definitions agree.

Proposition 1.5. Let $\underline{A}$ be the category of finitely generated abelian groups. Then $K_0(\underline{A}) = Z$.

Proof. Let $A \in \underline{A}$. Then $A \cong Z^n \oplus Z/d_1Z \oplus \dots \oplus Z/d_rZ$ by the fundamental theorem on finitely generated abelian groups. Hence

$$[A] = n[Z] + \sum_{i=1}^{r} [Z/_{d_i}Z] \text{ in } K_0(\underline{A}).$$

But the exact sequence $0 \longrightarrow Z \xrightarrow{\hat{d}} Z \longrightarrow Z/dZ \longrightarrow 0$ shows that $[Z/dZ] = 0$. Hence $[A] = n[Z]$, and $K_0(\underline{A})$ is cyclic. Let $rk: \underline{A} \longrightarrow Z$ by $rk(A)$ = number of free cyclic summands of A. Then if $0 \longrightarrow A' \longrightarrow A \longrightarrow A'' \longrightarrow 0$ is exact in $\underline{A}$, $rk\ A = rk\ A' + rk\ A''$. We conclude, as in proposition 1.2 that $K_0(\underline{A}) = Z$. Done.

Proposition 1.6. Let $\underline{A}$ be the category of finite abelian groups. Then $K_0(\underline{A})$ is free abelian with base $[Z/pZ]$, p running over the primes.

Proof. Any $A \in \underline{A}$ has a composition series $A = A_n \supset A_{n-1} \supset \dots \supset A_0 = 0$ where A_i/A_{i-1} is isomorphic to Z/p_iZ where p_i is a prime number.

We claim $[A] = \sum_{i=1}^{n} [A_i/A_{i-1}]$.

This is clear if $n = 1$. We induct on n. Hence we can assume $[A_{n-1}] = \sum_{1}^{n-1} [A_i/A_{i-1}]$. The exact sequence

$$0 \longrightarrow A_{n-1} \longrightarrow A_n \longrightarrow A_n/A_{n-1} \longrightarrow 0$$

shows that

$$[A_n] = [A_{n-1}] + [A_n/A_{n-1}].$$

Thus, the claim is established, and $[Z/pZ]$ generates $K_0(\underline{A})$. Let r_p: obj $\underline{A} \longrightarrow Z$ by $r_p(A)$ = number of A_i/A_{i-1} isomorphic to Z/pZ in any composition series for A. This is well defined by the Jordan-Hölder Theorem. r_p factors through $K_0(A)$. Hence, r_p induces a homomorphism

$$r_p: K_0(\underline{A}) \longrightarrow Z$$

and $r_p[Z/pZ] = {}_{pq}$ where q is any prime number. If $\sum_{i=1}^{r} a_{p_i}[Z/p_i Z] = 0$ in $K_0(\underline{A})$, then $r_q(\sum_{i=1}^{r} a_{p_i}[Z/p_i Z]) = 0$ implies $a_q = 0$. Hence $K_0(\underline{A})$ is free on the group generated by $[Z/pZ]$ where p is prime. Done.

<u>Definition</u>. Let $\underline{A}$ be an abelian category and $A \in \underline{A}$. We say A is <u>simple</u> if it has no proper $\underline{A}$ subobjects. A has <u>finite length</u> if there exists a composition series

$$* \qquad A = A_n \supset A_{n-1} \supset \dots \supset A_0 = 0 \qquad \text{of}$$

subobjects of A with each A_i/A_{i-1} simple.

<u>Theorem 1.7</u>. If every object of $\underline{A}$ has finite length, then $K_0(\underline{A})$ is free abelian on $[A]$ where A runs over the set of isomorphism classes of simple objects of $\underline{A}$.

<u>Proof</u>. By applying the embedding theorem and the Jordan-Hölder theorem we know that the factors A_i/A_{i-1} in * are unique up to order and isomorphism. As in the claim in proposition 1.6.
$[A] = \sum [A_i/A_{i-1}]$. That is $\{[S]\}$ with S simple generate $K_0(\underline{A})$.

Let S be simple and $A \in \underline{A}$. Define $r_S(A)$ = the number of quotients isomorphic to S in any composition series for A. As before r_S is well defined and factors through $K_0(\underline{A})$. We call the induced map r_S. If B is simple then

$$r_S(B) = \begin{cases} 0 \text{ if A is not isomorphic to S} \\ 1 \text{ if A is isomorphic to S.} \end{cases}$$

As in proposition 1.6. the generators [S] are independent.

Done.

Now we consider a distinguished set S of short exact sequences in a small abelian category $\underline{A}$. We define $K_0(\underline{A}, S)$ to be the abelian group generated by [A] with $A \in \underline{A}$ and relations $[A] = [A'] + [A'']$ for each $0 \longrightarrow A' \longrightarrow A \longrightarrow A'' \longrightarrow 0$ in S.

If S = the set of split exact sequences of $\underline{A}$, then $K_0(\underline{A}, S)$ (also denoted by $K_0(\underline{A}, +)$) is characterized by the conditions:

1) $K_0(\underline{A}, +)$ is generated by [A] for each $A \in \underline{A}$,

2) $[A] = [A'] + [A'']$ if A is isomorphic to $A' \oplus A''$, and

3) A isomorphic to C implies $[A] = [C]$.

We can generalize this construction in the following manner:

Let $F: \underline{A} \times \underline{A} \longrightarrow \underline{A}$ be a bifunctor. We define $K_0(\underline{A}, F)$ to be the abelian group generated by all [A] for $A \in \underline{A}$ with relations

1) A isomorphic to B implies $[A] = [B]$ and

2) $[F(A, B)] = [A] + [B]$.

Then $K_0(A, +)$ is $K_0(\underline{A}, F)$ where $F(A, B) = A \oplus B$.

Let R be a commutative ring with unit. A finitely generated projective R module P has rank 1 if $R_p \otimes_R P$ is isomorphic to R_p for all prime ideals p of R. If P and Q are rank 1 R modules then $P \otimes_R Q$ is rank 1 also since

$$R_p \otimes_R (P \otimes_R Q) \cong (R_p \otimes_R P) \otimes_{R_p} (R_p \otimes_R Q) \cong$$
$$\cong R_p \otimes_{R_p} R_p \cong R_p \ .$$

Let $\underline{A}$ be the category of finitely generated R modules of rank 1.

Definition. $K_0(\underline{A}, \otimes_R)$ = Pic(R), the Picard group of R.

Theorem 1.8. Pic(R), as defined, is isomorphic to the classical Picard group.

Proof. We have to show the set of isomorphism classes of finitely generated projective rank 1 modules is a group under $\otimes_R$. The operation is clearly well defined and associative. We want that inverses exist. Let P be finitely generated, rank n, and projective. We claim $P^* = \mathrm{Hom}_R(P, R)$ is its inverse.

1) P^* is projective and finitely generated. Say $P \oplus Q \cong F$ with F finitely generated free. Then $P^* \oplus Q^* \cong F^* \cong \coprod_{i=1}^{n} R^* \cong \coprod_{1}^{n} R$.

2) P is rank 1.

$$\begin{aligned}(P^*)_p &= (\mathrm{Hom}_R(P, R))_p \\ &= \mathrm{Hom}_{R_p}(P_p, R_p) \\ &\cong \mathrm{Hom}_{R_p}(R_p, R_p) \\ &\cong R_p \qquad \text{for every prime ideal } p.\end{aligned}$$

3) $P \otimes P^* \cong R$

We map $P \otimes P^*$ to R by $x \otimes f \rightsquigarrow f(x)$. This is an isomorphism. We only need to check the map at each maximal ideal m of R. We have

$$\begin{array}{ccc} P_m \otimes_{R_m} (P^*)_m & \longrightarrow & R_m \\ \| \quad\quad \| & & \\ R_m \otimes_{R_m} R_m & \longrightarrow & R_m \ . \end{array}$$

Hence the set of isomorphism classes forms a group and is a solution to the same universal problem as Pic(R). Therefore they are isomorphic. Done.

The classical Brauer group can be obtained from consideration of K_0. Let k be a field and $M_n(k)$ be the $n \times n$ matrices over k. Then the finite dimensional central simple k algebras, A, B, ..., can be divided into equivalence classes by $A \sim B$ if and only if there exists m and n so that $A \otimes_k M_n(k)$ is isomorphic to $B \otimes_k M_m(k)$. It is easy to see that $A \sim B$ if and only if they have the same Wedderburn constituent division ring.

The Brauer group of k, Br(k), is the set of equivalence classes with multiplication defined by

$$[A]\cdot[B] = [A \otimes_k B].$$

The class of k is a unit and if A^O is opposite algebra for A (i.e., $A^O = A$ as vector space and ab in A^O means ba in A). Then

$$[A \otimes_k A^O] = [k]$$

and $[A^O]$ is the inverse of [A].

Let $\underline{A}$ be the category of finite dimensional central simple k algebras. The group $K_0(\underline{A}, \otimes_k)$ is universal for functions f: obj $\underline{A} \longrightarrow$ abelian groups such that

1) A isomorphic to B implies f(A) = f(B), and
2) $f(A \otimes_k B) = f(A) + f(B)$.

Let $G = Q^{+*}$, the multiplicative group of positive rational numbers. The map $\mathcal{E}$: obj $A \longrightarrow G$ given by $\mathcal{E}(A) = (\dim_k A)^{1/2}$ is an additive map and hence factors through $K_0(\underline{A}, \otimes_k)$. But G is free abelian on the generators {p} for p a prime number, and the map

$n\colon G \longrightarrow K_0(\underline{A}, \otimes_k)$ given by $n(p) = [M_p(k)]$ is a splitting for $\mathcal{E}$. That is $\mathcal{E}n = 1_G$. Let $\tilde{K}_0(\underline{A}, \otimes_k)$ be the kernel of $\mathcal{E}$. Then we have

$$K_0(\underline{A}, \otimes_k) = \tilde{K}_0(\underline{A}, \otimes_k) \oplus Q^{+*}.$$

Theorem 1.9. $\tilde{K}_0(\underline{A}, \otimes_k)$ is isomorphic to $B_r(k)$.

Proof. We only need to prove they both satisfy the same universal problem. By definition $\tilde{K}_0(\underline{A}, \otimes_k)$ is universal for maps $f\colon \text{obj } A \longrightarrow$ abelian groups satisfying:

1) A isomorphic to B implies $f(A) = f(B)$,

2) $f(A \otimes_k B) = f(A) + f(B)$, and

3) $f(M_p) = 0$.

But 1), 2), and 3) and the fact that $M_n(k) \otimes M_m(k)$ is isomorphic to $M_{mn}(k)$ proves that $f(M_n) = 0$ for all n. It is now clear that $\tilde{K}_0(\underline{A}, \otimes_k)$ is isomorphic to Br(k). Done.

Remark.1) Let $\underline{G}$ be the category of finite groups. Let K_0 be generated by the objects of $\underline{G}$ with relations G isomorphic to H implies [G] = [H] and if N is a normal subgroup of G then [G] = [N] + [G/N]. Then K_0 is free abelian on the classes of the simple groups including all the Z/pZ.

2) (Watts) Let I be the category of finitely triangulable spaces. K_0 has the relations: 1) A isomorphic to B implies [A] = [B] and 2) If (X, A) is a triangulable pair then [X] = [A] + [X/A]. Then K_0 is isomorphic to Z, and this isomorphism is given by the Euler characteristic.

Definition. If $\underline{A}$ is a small abelian category and $\underline{B}$ a full subcategory of $\underline{A}$ we define $K_0(\underline{B})$ to be the group generated by [B] for

$B \in$ obj $\underline{B}$ and relations: If $0 \to B' \to B \to B'' \to 0$ is exact in $\underline{A}$ and B, B', B" are all in obj B then

$$[B] = [B'] + [B''].$$

Remarks. 1) If $\underline{B} = \underline{A}$, then the two definitions agree.

2) $K_0(B)$ depends on $\underline{A}$, the containing abelian category.

3) If $\underline{P}$ is the full subcategory of projective objects of $\underline{A}$, it is clear that $K_0(\underline{P}) = K_0(\underline{P}, \oplus)$ since every short exact sequence splits.

Definition. Let R be any ring with unit and, ${}_R\underline{M}$ be the category of finitely generated left R modules, and ${}_R\underline{P}$ be the full subcategory of finitely generated projective left R modules. Then we define

$$\underline{K_0(R)} = K_0(\underline{P}) = K_0(\underline{P}, \oplus).$$

Remarks. 1) Let X be a connected compact Hausdorff space and R be the ring of continuous real valued functions. Then $K_0(R)$ is isomorphic to $K^0_{\mathbb{R}}(X)$, topological K of real vector bundles.

2) Let $\underline{A}$ be abelian with full subcategories $\underline{C} \subset \underline{B} \subset \underline{A}$. Then the inclusion of $\underline{C}$ into $\underline{B}$ induces a map $K_0(\underline{C}) \to K_0(\underline{B})$. If $\underline{P}$ is the full subcategory of projectives in $\underline{A}$, the category of finitely generated modules over a ring, then this map generalizes the classical Cartan matrix.

Theorem 1.10. Let A, B be objects in any abelian category $\underline{A}$. If $[A] = [B]$ in $K_0(\underline{A}, +)$, then there is a $C \in \underline{A}$ such that $A \oplus C$ is isomorphic to $B \oplus C$.

Proof. Let G be the free abelian group generated by isomorphism classes of objects in $\underline{A}$ and let H be the subgroup generated by all elements of G of the form $[A \oplus B] - [A] - [B]$. Then $K_0(\underline{A}, +) = G/H$. Hence in the group G, $[A] - [B] \in H$. That is

$$[A]-[B] = \sum_i([C_i \oplus D_i]-[C_i]-[D_i])-\sum_j([E_j \oplus F_j]-[E_j]-[F_j]) .$$

By rearranging the terms we get that

$$[A]+\sum[C_i]+\sum[D_i]+\sum[E_j \oplus F_j] = [B]+\sum[C_i \oplus D_i]+\sum[E_j]+\sum[F_j] .$$

in O. This means that the modules on the left hand side are isomorphic to those on the right in some order so the same equation holds for the direct sums. The theorem is proved by setting

$$C = \sum C_i \oplus \sum D_i \oplus \sum(E_j \oplus F_j).$$

Done.

Theorem 1.11. Let A and B be in an abelian category $\underline{A}$. Then $[A] = [B]$ in $K_0(\underline{A})$ if and only if there exist short exact sequences $0 \longrightarrow C' \longrightarrow C \longrightarrow C'' \longrightarrow 0$ and $0 \longrightarrow D' \longrightarrow D \longrightarrow D'' \longrightarrow 0$ in $\underline{A}$ such that $A \oplus C' \oplus C'' \oplus D$ is isomorphic to $B \oplus C \oplus D' \oplus D''$.

Proof. $\Leftarrow$ is obvious.

$\Rightarrow$ Let G be as above and let H be the subgroup generated by all elements of G of the form $[A] - [A'] - [A'']$ where there is an exact sequence $0 \longrightarrow A' \longrightarrow A \longrightarrow A'' \longrightarrow 0$ in $\underline{A}$. If $[A] = [B]$ in $K_0(\underline{A})$, then $[A] - [B] = \sum_i([C_i] - [C_i'] - [C_i'']) - \sum_j([D_j] - [D_j'] - [D_j''])$. Therefore $A \oplus \sum C_j' \oplus \sum C_i'' \oplus \sum D_j$ is isomorphic to $B \oplus \sum C_i \oplus \sum D_j' \oplus \sum D_j''$. By setting $C' = \sum C_i'$, $C = \sum C_i$, $C'' = \sum C_i''$, $D' = \sum D_j'$, $D = \sum D_j$, and $D'' = \sum D_j''$ we obtain the conclusion of the theorem.

Done.

Definition. Let $\underline{A}$ be an additive category and A, B, C, and D $\in \underline{A}$. We say (A, B) $\sim$ (C, D) if there exist E and F $\in \underline{A}$ such that $A \oplus E$ is isomorphic to $C \oplus F$ and $B \oplus E$ is isomorphic to $D \oplus F$.

Remark. This is an equivalence relation.

Let G be the set of equivalence classes. We define addition on G by $(A, B) + (C, D) = (A \oplus C, B \oplus D)$.

Proposition 1.12. G is a group which is isomorphic to $K_0(\underline{A}, +)$.

Proof. The zero element is represented by (0, 0). Since $(C, C) \sim (0, 0)$ for any $C \in \underline{A}$, the inverse of (A, B) is (B, A). Hence G is an abelian group. Let f: obj A $\longrightarrow$ G by $f(A) = [(A, 0)]$. f factors through $K_0(\underline{A}, +)$. Call the resulting map f. $g: G \longrightarrow K_0(\underline{A}, +)$ given by $g([(A, B)]) = [A] - [B]$ is clearly the inverse of f. Done.

Chapter 2. Krull-Schmidt Theorems and Applications

Definition. Let A be an object in an additive category $\underline{A}$. A is indecomposable if A isomorphic to $B \oplus C$ implies either B or C is the zero object.

Definition. Let $\underline{A}$ be an additive category. Then the Krull-Schmidt Theorem holds in $\underline{A}$ if

1) Every object is a finite direct sum of indecomposable objects, and
2) This decomposition is unique up to isomorphism. That is if $A_1 \oplus \ldots \oplus A_n$ is isomorphic to $B_1 \oplus \ldots \oplus B_m$ with all A_i and B_j non zero indecomposables, then $m = n$ and there is a permutation, σ, of $\{1, \ldots, n\}$ such that A_i is isomorphic to $B_{\sigma(i)}$ for all i.

Theorem 2.1. If $\underline{A}$ is an additive category for which the Krull-Schmidt theorem holds then

1) $K_0(\underline{A}, +)$ is a free abelian group on $\{[A]\}$ where A runs over a set of isomorphism classes of indecomposible objects of $\underline{A}$.

2) If $[A] = [B]$ in $K_0(\underline{A}, +)$, then A is isomorphic to B.

Proof. Let $A \in \underline{A}$. Then A is isomorphic to $A_1 \oplus \ldots \oplus A_n$ with each A_i indecomposable. Define $n_j(A)$ = the number of times A_j occurs in the direct sum. As before the map which sends [A] to $\sum n_j(A)[A_j]$ is the desired isomorphism. Done.

Definitions. Let $A \in \underline{A}$ where $\underline{A}$ is an abelian category. Then we say A has the ascending chain condition (A.C.C.) if

$A \xrightarrow[p_1]{} A_1 \xrightarrow[p_2]{} A_2 \xrightarrow[p_3]{} A_3 \longrightarrow \cdots$ is a sequence of maps and objects in $\underline{A}$ with each p_i an epi then there exists an n_0 such that if $n \geq n_0$ then p_n is an iso. We say $\underline{A}$ has A.C.C. if every object has A.C.C. We say that A has the descending chain condition (D.C.C.) if

$\cdots \longrightarrow A_3 \xrightarrow[i_3]{} A_2 \xrightarrow[i_2]{} A_1 \xrightarrow[i_1]{} A$ is a sequence of maps and objects in $\underline{A}$ with each i_j a mono then there exists an n_0 such that if $n \geq n_0$ then i_n is an iso. We say that $\underline{A}$ has D.C.C. if every object of $\underline{A}$ has D.C.C. We say that A has the bichain condition (B.C.C.) if

$$A \underset{i_1}{\overset{p_1}{\rightleftarrows}} A_2 \underset{i_2}{\overset{p_2}{\rightleftarrows}} A_3 \underset{i_3}{\overset{p_3}{\rightleftarrows}} A_4 \rightleftarrows \cdots$$

is pair of sequences of maps and objects in $\underline{A}$ with all p_j epis and all i_j monos then there exists an n_0 such that if $n \geq n_0$ then p_n and i_n are both isos. We say $\underline{A}$ has B.C.C. if every object of $\underline{A}$ has B.C.C.

Remark. If A ($\underline{A}$) has A.C.C. and D.C.C. then A ($\underline{A}$) has B.C.C. The example of coherent sheaves over a complete algebraic variety shows he converse is false. In order to include this case, Atiyah has generalized the classical Krull-Schmidt theorem as follows.

Theorem 2.2. If A has B.C.C., then the Krull-Schmidt theorem holds in A.

Proof.

Lemma 2.3. If A has B.C.C. then A is a finite direct sum of indecomposables.

Proof. Clear.

Hence to prove the theorem it is enough to show this decomposition is unique. We will prove this in a series of lemmas and theorems.

Definition. A ring R is quasi-local if the non units in R form an ideal.

Proposition 2.4. A ring R is quasi-local if and only if x and y non units implies x + y is a non unit.

Proof. $\Longrightarrow$ obvious.

$\Longleftarrow$ It is enough to show that rx a unit implies x is a unit. This is clear if R is commutative. Otherwise let w be such that wrx = 1. Then xwr is an idempotent for (xwr)(xwr) = x(wrx)wr = xwr. Let xwr = e. e + (1 - e) = 1. If e is a unit in R, then $e^2 = e$ implies e = 1. Then x is a unit. If e is not a unit, then 1 - e must be a unit. Hence 1 - e = 1 and e = 0. But (wr)ex = (wr)(xwr)x = (wrx)(wrx) = 1. Hence e $\neq$ 0. Done with proposition 2.4.

Lemma 2.5. If A has B.C.C. and is indecomposable then End(A) is quasi-local. If End(A) is quasi-local, then A is indecomposable.

Proof. Let f and g $\in$ End(A) with f + g = 1_A. By proposition 2.4 it is enough to show that either f or g is an isomorphism. Let A_n be the image of $f^n: A \longrightarrow A$. Then $A_n \overset{f/A_n}{\underset{\text{inclusion}}{\rightleftarrows}} A_{n+1}$ defines a bichain. Hence there exists an n_0 such that for $n \geq n_0$, $f|A_n: A_n \longrightarrow A_{n+1}$ is an isomorphism. Define $\rho: A \longrightarrow A$ by $\rho = [(f|_{A_n})^n]^{-1} f^n$. Then $\rho | A_n = 1_{A_n}$ and im $\rho = A_n$. Hence, $A = A_n \oplus \ker \rho$. Either $A_n = A$ and f is an iso or $A_n = 0$ and $f^n = 0$. But then $g = 1 - f$ has as inverse $1 + f + f^2 + \ldots + f^{n-1}$ and g is an iso.

The second part of the lemma is obvious since any proper direct sum decomposition of A will yield f and $g \in \mathrm{End}(A)$ with $f + g = 1$.

Done with lemma 2.5.

Theorem 2.6. Let $\underline{A}$ be an abelian category and $A \in \underline{A}$. If End(A) is quasi local, then $A \oplus B$ isomorphic to $A' \oplus C$ and A' isomorphic to A implies B is isomorphic to C.

Proof. First a short lemma.

Lemma 2.7. (X-lemma). If the following diagram has exact rows and if the composite hf is an isomorphism, then gj is an isomorphism:

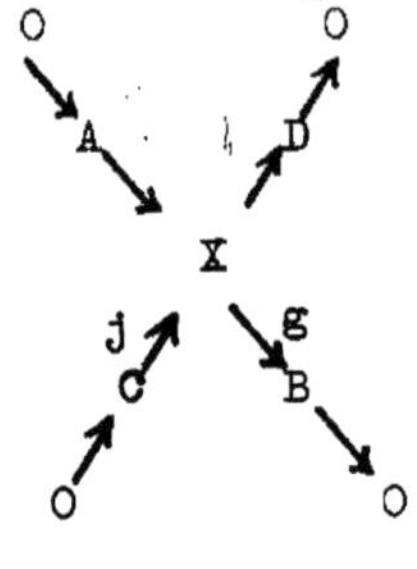

Proof. Apply the embedding theorem and check it for modules.

Done.

Back to theorem 2.6. By using the isomorphism of $A \oplus C$ with $A' \oplus B$ we obtain maps $A \longrightarrow A' \longrightarrow A$ and $A \longrightarrow C \longrightarrow A$ with sum the identity. Hence one of the two is an iso. If $A \longrightarrow A' \longrightarrow A$ is an iso, then let X be $A \oplus B$. We have

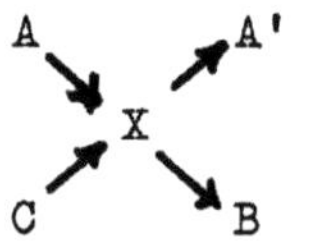

Applying lemma 2.7 gives that the map from C to B is an iso and we are done.

Suppose $A \xrightarrow{f} C \xrightarrow{g} A$ is an isomorphism. Let C' be the image of A in C. Then C' is a direct summand of C where the splitting of the inclusion is the map $f(gf)^{-1}g$. Let $C = C' \oplus C''$ with C' isomorphic to A. Consider the diagram

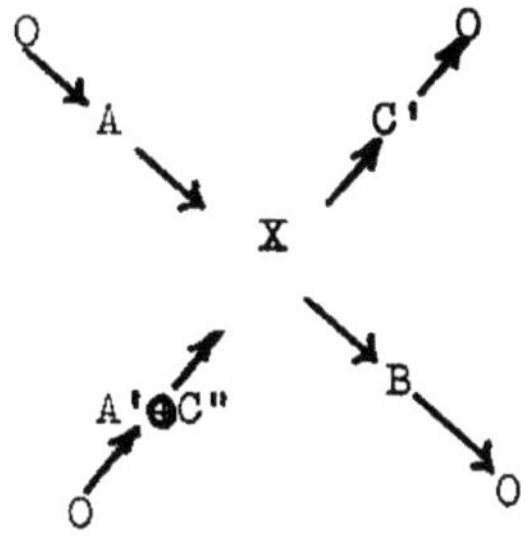

This yields $A' \oplus C''$ isomorphic to B. Since C' is isomorphic to A' we have C is isomorphic to B. Done with theorem 2.6

<u>Theorem 2.8</u>. Let $\underline{A}$ be an abelian category and A_i objects of $\underline{A}$ such that $\text{End}(A_i)$ are all quasi-local. If $A_1 \oplus \dots \oplus A_n$ is isomorphic to $B_1 \oplus \dots \oplus B_m$ and the B_j are all indecomposable, then $m = n$ and there exists a permutation σ of $\{1, \dots, n\}$ such that B_i is isomorphic to $A_{\sigma i}$ for all i.

<u>Remarks</u>. 1) This will complete the proof of theorem 2.2.

2) Theorem 2.8 is true for infinite direct sums if A has exact directed $\varinjlim$.

3) This almost proves that projective modules over quasi local rings are free.

Proof. First we prove a lemma.

Lemma 2.9. Let A_i be as in the above theorem. Suppose $X = \sum_{1}^{n} A_i$ and X is isomorphic to $B \oplus C$ where $B \neq 0$. Then there exists an i such that $A_i \longrightarrow B \longrightarrow A_i$ is an iso. If has exact directed $\varinjlim$, then the same is true for $n = \infty$.

Proof. There is a finite set, S, such that the map $B \longrightarrow \sum_{i \in I} A_i / \sum_{i \in S} A_i$ is not a monomorphism. This is obvious if n is finite. Let n be infinite and suppose there is no such set. Therefore the map $B \longrightarrow \varinjlim \sum_{i=1}^{\infty} A_i / \sum_{i \in S} A_i$ is a mono where $\varinjlim$ is taken over finite subsets of I. But the limit is 0. This contradicts the assumption if B is non zero.

The proof of the lemma proceeds by induction on the number of elements of S. Choose any $j \in S$. If $A_j \longrightarrow B \longrightarrow A_j$ is an iso we are done. If not, then $A_j \longrightarrow C \longrightarrow A_j$ is an iso since $\mathrm{End}(A_j)$ is quasi-local. Then $C = C' \oplus C''$ where C' is the image of A_j. Hence $A_j \longrightarrow C'$ is an iso. Apply the X lemma to the following diagram:

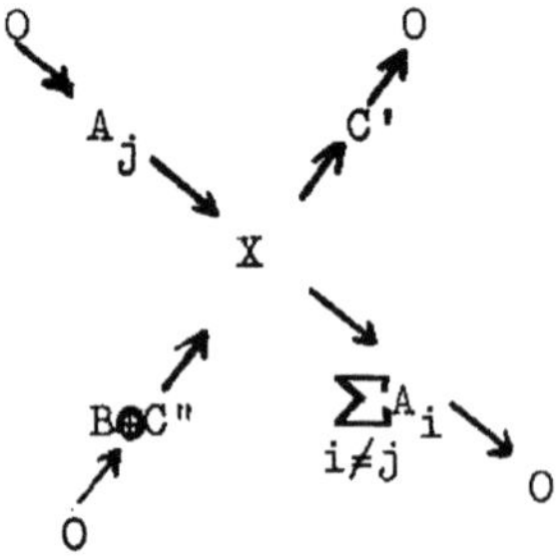

to obtain

$B \oplus C''$ is isomorphic to $\sum_{i \neq j} A_i$. Then $B \longrightarrow \sum_{i \neq j} A_i / \sum_{i \in S-\{j\}} A_i$ is not a mono.

If the number of elements in S is one we have a contradiction. If the number of elements in S is greater than 1, we have by induction that there is an $i \in S - \{j\}$ such that $A_i \longrightarrow B \longrightarrow A_i$ is an isomorphism. Done with lemma 2.9.

Corollary 2.10. Same hypothesis as in 2.9 but add that B is indecomposable and conclude that B is isomorphic to A_i for some i.

Back to theorem 2.8. Each B_j is isomorphic to some A_i by the corollary; and, therefore, $End(B_i)$ is quasi local. If n is finite, we use induction on n. Then A_1 is isomorphic to B_j for some j. By theorem 2.6. we conclude $\sum_{i=2}^{n} A_i$ is isomorphic to $\sum_{i \neq j} B_i$.

If n is infinite we want to show that if M is an indecomposable object then the number of A_i isomorphic to M equals the number of B_j with B_j isomorphic to M. If the number of such A_i is finite this is clear by the same argument as for finite N. Suppose it is infinite

For each i let $S_i = \{j \mid A_i \longrightarrow B_j \longrightarrow A_i \text{ is an iso}\}$. S_i is a finite set since there is a finite set S such that $A_i \longrightarrow \sum_j B_j / \sum_{j \in S} B_j$ is not a mono and S_i is clearly contained in S. Since B_j is indecomposable, $j \in S_i$ implies B_j is isomorphic to A_i. Since there is at least one A_i such that $A_i \longrightarrow B_j \longrightarrow A_i$ is an iso, we have that

$$\bigcup_{\{i \mid A_i \text{ is isomorphic to } M\}} S_i = \{j \mid B_j \text{ is isomorphic to } M\}.$$

Therefore the number of j such that B_j is isomorphic to M is $\leq$ the number of i such that A_i is isomorphic to M. But $End(B_j)$ is quasi local. Hence, by symmetry, we get $\geq$ and hence equality.

Done with theorems 2.2. and 2.8.

Theorem 2.11. Let $\underline{A}$ be an abelian category such that the following two conditions hold:

1) Each object of $\underline{A}$ is a finite direct sum of indecomposable objects.

2) A indecomposable implies End(A) is quasi local.

Then the Krull-Schmidt theorem holds for $\underline{A}$.

Proof. Clear.

Theorem 2.12. Let $\underline{A}$ be an abelian category and $A_i \in \underline{A}$ such that End(A) is quasi local. If $\coprod_{i=1}^{n} A_i = B \oplus C$ with $n < \infty$, then there is a finite set S such that $X = B \oplus \coprod_{i \in S} A_i$ is direct.

Proof. Let $X = B \oplus C$. Pick S maximal so that $B \oplus \coprod_{i \in S} A_i$ is direct and a direct summand of X. By replacing B by $B \oplus \sum_{i \in S} A_i$ we can assume that there is no i such that $B + A_i$ is direct and a direct summand of X. We want that $B = X$. Assume $B \neq X$. Then $B \oplus C = X$ and $C \neq 0$. Then there is an i such that $A_i \longrightarrow C \longrightarrow A_i$ is an isomorphism since $End(A_i)$ are quasi local for all i. Let $C' = im[A_i \longrightarrow C]$. Then $C = C' \oplus C''$. But then $X = A_i \oplus B \oplus C''$ since the map $A_i \longrightarrow X/B + C'' = C'$ is onto and $X \in A_i \cap B$ implies $X = 0$. This contradicts the assumption on B. Done.

Corollary 2.13. Same hypothesis as above then C is isomorphic to $\coprod_{i \in S} A_i$.

Corollary 2.14. If R is a quasi local ring then finitely generated projectives are free.

<u>Proof</u>. End(R) = Hom(R, R) = R^0; and, therefore, is quasi local. P a finitely generated projective implies there exist a finitely generated projective Q such that $P \oplus Q$ is free. Corollary 2.13 then implies P is free.

<u>Remark</u>. Kaplansky has shown that if R is quasi local, then all projectives are free.

<u>Proposition 2.15</u>. Let $\underline{A}$ be an abelian category. Suppose that for each $A \in \underline{A}$ there exists a 2 sided ideal I of Λ = End(A) such that

1) Every $x \in \Lambda$ which is a unit mod I is a unit,

2) Every idempotent of Λ/I lifts to an idempotent of Λ, and

3) Λ/I is semi-simple (With D.C.C.).

Then the Krull-Schmidt theorem holds in $\underline{A}$.

<u>Remark</u>. 1) and 3) imply that I is the Jacobson radical of Λ. (cf. Prop. 2.16.)

<u>Proof</u>. We will use conditions 1) and 2) of theorem 2.11. Suppose $A \in \underline{A}$ is not a finite direct sum of indecomposables. Then

$$A = A_i \oplus A_1'.$$
$$A_1 = A_2 \oplus A_2'$$
$$\cdots\cdots\cdots\cdots$$
$$A_n = A_{n+1} \oplus A_{n+1}'$$

with all A_i non zero. Let e_n be the natural projection of A onto A_n. Then the e_i clearly satisfy

a) $e_n^2 = e_n$,

b) $e_{n+1}e_n = e_ne_{n+1} = e_{n+1}$, and

c) $e_{n+1} \neq e_n$.

By b) $e_n - e_{n+1}$ is also idempotent.

Let $\bar{e}_n$ = the image of e_n in $\bar{\Lambda} = \Lambda/I$. Clearly a) and b) are satisfied in $\bar{\Lambda}$. If $e \in \Lambda$ is idempotent and $e \equiv 1(I)$ then 1) implies e is a unit and hence $e = 1$. If $e \equiv 0(I)$ then $1 - e$ is idempotent and a unit. Hence, $e = 0$. Hence, $\bar{e}_n = \bar{e}_{n+1}$ implies $e_n - e_{n+1}$ is idempotent and $e_n - e_{n+1} \equiv 0$ (I). Therefore $e_n - e_{n+1} \equiv 0$ contradicting c). Hence c) holds in $\bar{\Lambda}$.

Since $\bar{\Lambda}$ has D.C.C., the chain of ideals $\bar{\Lambda} \supset \bar{\Lambda}\,\bar{e}_1 \supset \ldots . \supset \bar{\Lambda}\,\bar{e}_n \supset$ must stop. $\bar{\Lambda}\,\bar{e}_n = \bar{\Lambda}\,\bar{e}_{n+1}$ for some n. Then $\bar{e}_n = a\bar{e}_{n+1}$. But $\bar{e}_{n+1} = \bar{e}_n\bar{e}_{n+1} = a\,\bar{e}^2_{n+1} = a\,\bar{e}_{n+1} = \bar{e}_n$. This is a contradiction. Hence we have verified 1) of theorem 2.11.

Let $A \in \underline{A}$ be indecomposible and $\Lambda = \mathrm{End}(A)$. If $e \in \Lambda$ is idempotent, then A is isomorphic to ker $e \oplus$ im e. But A indecomposible implies e equals 0 or 1. Hence Λ/I has no idempotents except 0 and 1.

By Wedderburn's theorem Λ/I is isomorphic to a direct sum of matrix rings over division rings. But since Λ/I has only 0 and 1 as idempotents, Λ/I must be a division ring.

Let $x \in \Lambda - I$. Then $\bar{x} \in \Lambda/I$ is non zero, and, hence, a unit. By 1) x is a unit. Hence I is the set of non units of Λ and Λ is a quasi local ring. Thus, 2) of theorem 2.11. is verified. Done.

<u>Proposition 2.16</u>. Let R be a ring and I a 2 sided ideal. Then the following are equivalent:

1) x a unit mod I implies x is a unit,

2) $x \equiv 1$ (I) implies x is a unit, and

3) I is contained in the Jacobson radical of R, J(R).

<u>Proof</u>. 2) $\Longrightarrow$ 1). If $\bar{x}$ is a unit, then there exists $\bar{y}$ such that $\overline{yx} = \overline{xy} = 1$. Then $yx \equiv 1$ (I) and $xy \equiv 1$ (I). So xy and yx are

units. Hence there exist u and v such that $xyv = uyx = 1$. But then $yv = uy$ since if both left and right inverses exist they are equal. Hence, x is invertible.

1) $\Longrightarrow$ 3) $J(R) = \bigcap M$ where M runs over the set of maximal ideals of R. For such M, $M + I = R$ or M. If $M + I = M$, then $I \subset M$. If $M + I = R$ then $1 = m + i$ for some $m \in M$ and $i \in I$. Then $\bar{m}$ is a unit. This implies m is a unit. But M is a proper ideal. Hence, $I \subset M$ for all maximal ideals M and $I \subset J(R)$.

3) $\Longrightarrow$ 2) Let $x \equiv 1\ (I)$. Then $Rx = R$. If $Rx \subset M$ where M is a maximal left ideal of R. Then $M \supset I$ and $x - 1 \in I$ implies $x - 1 \in M$. But $x \in M$ so $1 \in M$ which is a contradiction.

Since $Rx = R$, x has a left inverse y. Then $1 = yx$ and $yx \equiv y(I)$ implies $y \equiv 1(I)$. The same argument on y shows there exists a z with $zy = 1$. Hence x is a unit. Done.

If $I \subset R$ is a nilpotent 2 sided ideal, then 2) is easily verified. For if $x \equiv 1(I)$, then $x = 1 + a$ with $a \in I$. Say $a^n = 0$. The element $1 - a + a^2 - a^3 + \ldots + (-1)^{n-1}a^{n-1}$ is an inverse for x.

Proposition 2.17. If I is a nilpotent 2 sided ideal of R, then every idempotent of R/I lifts to one of R.

Proof. Suppose $I^n = 0$. We use induction on n. The sequence

$$R \longrightarrow R/I^{n-1} \longrightarrow R/I^{n-2} \longrightarrow \ldots \longrightarrow R/I^2 \longrightarrow R/I$$

will obey lifting property at each stage if we can prove the proposition for $n = 2$. So we assume $I^2 = 0$ and $e \in R/I$ is an idempotent.

Let $x \longrightarrow e$. Then $a = x^2 - x \in I$. Let $y = x + a(1 - 2x)$. Then $y \equiv x\ (I)$ and $y^2 - y = x^2 + 2xa(1 - 2x) + (\text{elt of } I^2) - x - a(1 - 2x)$.

Hence $y^2 - y = (x^2 - x) + 2xa - 4x^2a - a + 2xa = 4xa - 4x^2a =$

$$= 4(x - x^2)a = -4a^2 = 0.$$

So y is idempotent and $y \longrightarrow e$. Done.

Now if R has D.C.C. on left ideals, then R has a largest nilpotent ideal, I, which is equal to the radical of R. R/I is semi-simple with D.C.C. Proposition 2.15. and the preceeding remark give:

Theorem 2.18. Let $\underline{A}$ be an abelian category. If for each $A \in \underline{A}$ End(A) has D.C.C. on left (or right) ideals, then the Krull-Schmidt theorem holds in $\underline{A}$.

Example. Let X be a complete algebraic variety and $\underline{A}$ be the category of coherent sheaves on X. Then Hom(F, G) is a finite dimensional vector space for any F and $G \in \underline{A}$. The Krull-Schmidt theorem holds in $\underline{A}$.

Definition. Suppose R is a ring and I a 2 sided ideal. We say that R is I-complete if the natural map $R \longrightarrow \varprojlim R/I^n$ is an iso. Equivalently, R is I-complete if, in the topology on R determined by letting the I^n be a base of open neighborhoods of 0, every Cauchy sequence converges to a unique limit.

Proposition 2.19. Let I be a 2 sided ideal of R such that R is I complete. Then

1) Every unit mod I is a unit of R.
2) Every idempotent of R/I lifts to one of R.

Proof. Consider the sequence of projections

$$R \longrightarrow \dots \longrightarrow R/I^{n+1} \longrightarrow R/I^n \longrightarrow \dots \longrightarrow R/I \longrightarrow 0.$$

Let $x \in R$ such that $\bar{x}_1$, the image of x in R/I, is a unit. Let $\bar{x}_n$ be the image in R/I^n. By proposition 2.16 $\bar{x}_n$ is a unit in R/I^n.

Let $\longrightarrow y_n \longrightarrow y_{n-1} \longrightarrow \dots \longrightarrow y_1$ be the sequence of inverses and let $y = \lim y_i$. By completeness $y \in R$, and yx and xy have image 1 in every R/I^n. Therefore $xy = yx = 1$ and x is a unit.

If $e_1 \in R/I$ is idempotent, we can lift it back one stage at a time to get e_n idempotents in R/I^n such that $e_n \longrightarrow e_{n-1}$ in the natural projection of $R/I^n \longrightarrow R/I^{n-1}$ by proposition 2.17. Let $e = \lim e_n$. Then $e^2 - e = 0$ since $e_n^2 - e_n = 0$ for all n. Thus e is an idempotent with image e_1 in R/I. Done.

<u>Theorem 2.20</u>. Let $\underline{A}$ be an abelian category such that for each $A \in \underline{A}$ there exists a 2 sided ideal I of R = End(A) such that R is I complete and R/I has left or right D.C.C. Then the Krull Schmidt theorem holds in $\underline{A}$.

<u>Proof</u>. Let J = the ideal of R such that R/J is isomorphic to $R/I/J(R/I)$ where J(R/I) is the Jacobson radical of R/I. Then 1) any unit mod J is a unit, and 2) any idempotent of R/J lifts to one of R since the sequence $R \longrightarrow R/I \longrightarrow R/J$ allows lifting in each stage because J/I is a nilpotent ideal of R/I. R/J is semi simple (with D.C.C.). Therefore, proposition 2.15 completes the proof.

<u>Lemma 2.21</u>. Let R be a commutative ring and A and B R modules such that A is finitely generated and B is noetherian. Then $A \otimes_R B$ and $\mathrm{Hom}_R(A, B)$ are noetherian.

<u>Proof</u>. We can find a finitely generated free module R^n and an epi $R \longrightarrow A \longrightarrow 0$. Tensoring with B gives

$$R^n \otimes_R B \longrightarrow A \otimes_R B \longrightarrow 0 \text{ is exact.}$$

Hom with B gives

$$0 \longrightarrow \mathrm{Hom}_R(A,\ B) \longrightarrow \mathrm{Hom}_R(R^n,\ B) \text{ is exact.}$$

$A^n \otimes_R B$ is isomorphic to $\sum_1^n B$, and $\mathrm{Hom}_R(R^n,\ B)$ is isomorphic to $\sum_1^n B$. But subobjects and quotients of noetherian objects are noetherian.

Done.

<u>Corollary 2.22</u>. Let R be a commutative, noetherian local ring which is complete with respect to its maximal ideal M. Let A be an R-algebra and <u>A</u> the category of A modules which are finitely generated as R modules. Then the Krull-Schmidt theorem holds in <u>A</u>.

<u>Proof</u>. Let $E = \mathrm{Hom}_A(B,\ B)$ for $B \in$ <u>A</u>. Then $E \subset \mathrm{Hom}_R(B,\ B)$ which is finitely generated by lemma 2.21. Therefore, E/ME is a finite dimensional algebra over R/M. Hence, E/ME has D.C.C.

But the Artin-Rees theorem implies every finitely generated module over a complete local ring is complete.

Done.

<u>Definition</u>. Let <u>A</u> be any category, A and B $\in$ <u>A</u>, and $f\colon A \longrightarrow B$. Then f is an <u>essential epi</u> if

1) f is an epi, and

2) If $X \xrightarrow{g} A$ is such that fg is an epi, then g is an epi.

<u>Remark</u>. The composition of essential epis is an essential epi.

<u>Definition</u>. $f\colon P \longrightarrow A$ is a <u>projective cover of A</u> is P is projective and f is an essential epi.

<u>Remark</u>. P is determined by A up to (a not necessarily unique) iso. For it P and Q are projective covers of A, we have the diagram

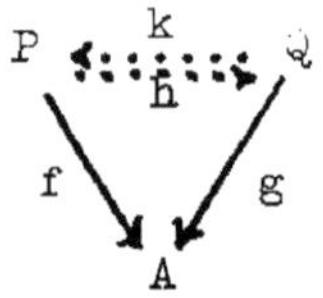

.

h exists since P is projective and is an epi since g is an essential epi. Q projective implies k exists such that $hk = 1_Q$. $g = ghk = fk$ and f essential implies k is an epi. $khk = k$ and k an epi implies $kh = 1_P$. Hence h and k are inverse isos.

Lemma 2.23. (Nakayama's Lemma). Let I be a 2 sided ideal of a ring R which is contained in the Jacobson radical of R. If M is a finitely generated R module such that $M/IM = 0$, then $M = 0$.

Proof. Well known.

Corollary 2.24. R and I as above. Let M be a finitely generated R module, then $M \longrightarrow M/IM$ is an essential epi.

Proof. Suppose $X \xrightarrow{g} M \xrightarrow{p} M/IM$ is such that pg is an epi. If $N \subset M$ is the image of g we have that $N + IM = M$. But then $M/N = I(M/N)$. Hence $M/N = 0$ by Nakayama's Lemma and $N = M$. Thus g is an epi.

Done.

Corollary 2.25. R and I as above. Let P be a finitely generated projective R module. Then $P \longrightarrow P/IP$ is a projective cover.

Remark. Using the above we can give another proof of corollary 2.14. Let P be a finitely generated projective module over a quasi local ring R with maximal ideal M. Then $P \longrightarrow P/MP$ is a projective cover and P/MP is a projective R/M module. But R/M is a division ring. Therefore, P/MP is free on n generators. Let F be a free R module on n generators. Then $F \longrightarrow F/MF$ is a projective cover of P/MP via the isomorphism of P/MP and F/MF. Uniqueness of projective cover shows P is free on n generators.

Theorem 2.26. If I is a 2 sided ideal in R where R is I complete, then the correspondence $P \rightsquigarrow P/IP$ determines a 1 - 1 correspondence between isomorphism classes of finitely generated projective R modules and finitely generated projective R/I modules.

Proof. If P is a finitely generated projective R module then there exists an R module Q such that $P \oplus Q$ is a finitely generated free R module. Therefore

$$P/IP \oplus Q/IQ \cong F/IF \cong R/I \otimes_R F$$

which is a finitely generated free R/I module. Therefore P/IP is a finitely generated projective R/I module. If P/IP is isomorphic to Q/IQ with both P and Q finitely generated projective R modules, then P is isomorphic to Q by the uniqueness of projective covers.

We claim the inverse correspondence is given by the projective cover. Let $\overline{P}$ be a finitely generated projective R/I module. Then $\overline{P} \oplus \overline{Q}$ is isomorphic to $\overline{F}$ a finitely generated free R/I module. $\overline{F}$ is isomorphic to F/IF where F is a free R module on the same number of generators.

The projection $\overline{R}$: $\overline{F} \longrightarrow \overline{P} \longrightarrow \overline{F}$ determines an idempotent $n \times n$ matrix over R/I where n is the number of generators of $\overline{F}$.

We want to lift $\overline{e}$ back to an idempotent $n \times n$ matrix over R via the projection $M_n(R) \xrightarrow{p} M_n(R/I)$. p is onto and has kernel $M_n(I)$. Hence if $M_n(R)$ is $M_n(I)$ complete we can lift e.

Lemma 2.27. Let I and J be 2 sided ideals of R. Then $M_n(I)M_n(J) = M_n(IJ)$.

Proof. $\subset$ is clear.

$\supset$ the matrices with ij in the $k\ell$ position where $i \in I$, $j \in J$ and $0 \leq k, \ell \leq n$ and 0 in all other positions generate $M_n(IJ)$. But it is the product of the matrix with i in the $k\ell$ place and 0 elsewhere and the matrix with j in the ~~11~~ place and 0 elsewhere.

Done.

Hence $M_n(I^m) = (M_n(I))^m$. Hence

$$M_n(R) \longrightarrow \varprojlim M_n(R)/M_n(I^m) \cong \varprojlim M_n(R/I^m).$$

This is an iso since R is I complete. Hence $\bar{e}$ lifts to an idempotent $e \in M_n(R)$. Then $F = P \oplus Q$ where $P = \text{im } e$ and $Q = \ker e$. $F/IF \cong P/IP \oplus Q/IQ$ and the induced image of $e: F/IF \longrightarrow F/IF$ is P/IP. But e mod I induces $\bar{e}$. Therefore, $P/IP \cong \bar{P}$. Done.

Corollary 2.28. If R is complete with respect to a 2 sided ideal I and R/I has D.C.C., then every finitely generated R module has a projective cover which is finitely generated.

Proof. Assume R/I is semi simple. If A is a finitely generated R module, A/IA is a projective R/I module. Therefore $P/IP \cong A/IA$ where P is a projective cover of A/IA. Then there exist $h: P \longrightarrow A$ such that

$$\begin{array}{ccc} P & \overset{h}{\cdots\cdots\to} & A \\ \downarrow & & \downarrow \\ P/IP & \xrightarrow{\ \cong\ } & A/IA \end{array}$$

commutes.

h is an epi since P is a projective cover of P/IP. It is easy to see that h makes P a projective cover of A.

If R/I is not semi-simple, then R/I is J complete where $J = \text{rad}(R/I)$ since J is nilpotent. Hence the R/I module A/IA has a projective cover $\bar{P}$. Using the correspondence in theorem 2.26. $\bar{P}$ corresponds to a finitely generated projective R module P. P is a projective cover of A. Done.

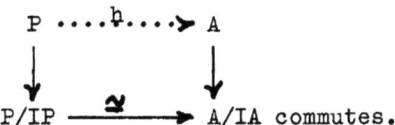

If $h: R \longrightarrow R'$ is a homomorphism of rings, then there is an induced map $K_0(h): K_0(R) \longrightarrow K_0(R')$ given by $[P] \longrightarrow [R' \otimes_R P]$. If P is a finitely generated projective R module, then there exists Q such that $P \oplus Q = F$ is a finitely generated free module. Then $(R' \otimes_R P) \oplus (R' \otimes_R Q) \cong R' \otimes_R F \cong$ free R' module. Hence $R' \otimes_R P$ is a finitely generated projective R' module. $K_0(h)$ is clearly additive.

Proposition 2.29. If R is complete with respect to a 2 sided ideal I, the map $K_0(R) \longrightarrow K_0(R/I)$ induced by the projection $R \longrightarrow R/I$ is an iso.

Proof. The inverse map is $[\bar{P}] \rightsquigarrow [P]$ when $P/IP = R/J \otimes_R P$ as in theorem 2.26.

Chapter 3. Definition of G(R) and Examples

Theorem 3.1. (A. Heller). Let $\underline{A}$ be an abelian category, $\underline{C}$ and $\underline{B}$ be full subcategories with $\underline{C} \subset \underline{B}$, and such that

1) $\underline{C}$ is closed in $\underline{A}$ under subobjects and quotient objects, and
2) Every object of $\underline{B}$ has a finite filtration with all quotients in $\underline{C}$.

Then the cannonical map $K_0(\underline{C}) \longrightarrow K_0(\underline{B})$ is an isomorphism.

Proof. We define an inverse map, f, from $K_0(\underline{B})$ to $K_0(\underline{C})$ as follows: let $B \in \underline{B}$. By 2) we pick a filtration $B = B_0 \supseteq B_1 \supset \ldots \supseteq B_n = 0$ so that each $B_i/B_{i+1} \in \underline{C}$. Then f(B) is defined to be $\sum_{i=0}^{n-1} [B_i/B_{i+1}]$. To show that this is independent of choice of filtration we first remark that the Jordan-Holder theorem (any two filtrations have a common refinement) proves we only need to check for refinements.

Induction proves it is enough to check for one insertion. Say $B_i \gtrdot B_{i+1}$ is changed to $B_i \gtrdot B' \gtrdot B_{i+1}$. Then B_i/B_{i+1} is replaced by B_i/B' and B'/B_{i+1}. First we need that B_i/B' and B'/B_{i+1} are in $\underline{C}$. We have the exact sequence

$$* \qquad 0 \longrightarrow B'/B_{i+1} \longrightarrow B_i/B_{i+1} \longrightarrow B_i/B' \longrightarrow 0$$

in $\underline{A}$ with $B_i/B_{i+1} \in \underline{C}$. But $\underline{C}$ is closed under quotients and subobjects. Therefore, B'/B_{i+1} and $B_i/B' \in \underline{C}$.

$*$ also proves that $[B_i/B_{i+1}] = [B_i/B'] + [B'/B_{i+1}]$

in $K_0(\underline{C})$. Therefore, the refinement gives the same element in $K_0(\underline{C})$. Hence f is well defined. It is clear that f is an iso. Done.

Corollary 3.2. Let $\underline{A}$ be an abelian category in which each object has finite length. Then $K_0(\underline{A})$ is the free abelian group on $[S]$ where S runs over a set of representatives of the simple objects of $\underline{A}$.

Proof. Let $\underline{B}$ be $\underline{A}$ and $\underline{C}$ be the full subcategory of simple objects in Theorem 3.1. Done.

Definitions. We say an object A of an abelian category is noetherian if A has A.C.C. We say a collection $\underline{S}$ of subobjects of A has sup $\underline{S}$ = A if whenever B is a subobject of A and $S \leqslant B$ for all $S \in \underline{S}$ then B = A. We say A is finitely generated if for every collection $\underline{S}$ of subobjects of A with sup $\underline{S}$ = A there is a finite subset $\underline{S}_0 \subseteq \underline{S}$ such that sup $\underline{S}_0$ = A.

Remark. If A is noetherian then every subobject is finitely generated. If $\underline{A}$ is countably right complete, then the converse is true.

Proposition 3.3. Let $\underline{A}$ be an abelian category. The noetherian objects of $\underline{A}$ form a Serre subcategory.

Proof. Let $0 \longrightarrow A' \xrightarrow{i} A \xrightarrow{j} A'' \longrightarrow 0$ be exact. That A noetherian implies A' and A" are is clear. Let A' and A" be noetherian and $A_1 \subsetneq A_2 \subsetneq A_3 \subsetneq \ldots$ be a chain of subobjects of A. Then there is an n_0 such that for $n \geq n$. $j(A_n)$ equals $j(A_{n+1})$ and $i^{-1}(A_n)$ equals $i^{-1}(A_{n+1})$. The five lemma shows A_n equals A_{n+1}. Done.

Definition. Let R be any ring. Then $G_0(R)$ is $K_0(\underline{A})$ where $\underline{A}$ is the category of noetherian left R modules.

Remark. If R is a left noetherian ring then $\underline{A}$ is the category of finitely generated left R modules.

Theorem 3.4. Let I be a nilpotent 2 sided ideal of a ring R then the map $G_0(R/I) \longrightarrow G_0(R)$ given by $[A] \rightsquigarrow [A]$ is isomorphism.

Proof. Let $\underline{A}$ be the category of all R modules, $\underline{B}$ be the category of all noetherian R modules, and $\underline{C}$ be the category of all objects C, of $\underline{B}$ such that IC = 0 (that is all noetherian R/I modules). Condition 1) of Heller's theorem is clear. Let $B \in \underline{B}$ and $I^n = 0$, then $B \supsetneq IB \supsetneq I^2B \supsetneq \ldots \supsetneq I^nB = 0$ is the filtration required for condition 2). Done.

Example. This example shows that the above theorem is false with finitely generated replacing noetherian.

Let k be a field, let J be the ideal of $k[x_1, x_2, \ldots, x_n, \ldots]$ generated by elements of the form x_ix_j. Let $R = k[x_1, x_2, \ldots, x_n, \ldots]/J$. Let I be the ideal generated by the images of the x_i's. Then $I^2 = 0$ and R/I is isomorphic to k. Consider the exact sequence $0 \longrightarrow I \longrightarrow R \longrightarrow k \longrightarrow 0$. Since I is not finitely generated, the category of finitely generated modules is not closed with respect to kernels, and, therefore, is not abelian.

In spite of this one still might hope that K_0 (category of finitely generated R modules) was isomorphic to K_0 (category of finitely generated R/I modules). R/I is a field and, therefore, $G_0(R/I) = K_0$ (category of finitely generated R/I modules) = Z. However, K_0 (category of finitely generated R modules) = 0.

It is certainly generated by [R/J] where J runs over the ideals of R. Therefore, it is enough to show there are all 0.

<u>Case 1</u>. $\dim_k J = \infty$. Let $\{j_n\}$ be a k basis for J. $j_n = \sum a_{nk}e_k$, where e_k is the image of x_k, since $J \subset I$. Let M be a vector space over k with basis a, b, c_1, c_2, ... We make M into a finitely generated R module by defining $e_i a = c_i$, $e_i b = \sum a_{ik}c_k$, and $e_i c_k = 0$. We can inject R into M in two ways. If we send 1 to a we get a map with coker k. If we send 1 to b we get a map with coker R/J. Therefore [R/J] = [k]. Let J be the ideal generated by $\{e_2, e_3, \ldots\}$. Then mapping k to R/J by sending 1 to image of e_1 gives the exact sequence

$$0 \longrightarrow k \longrightarrow R/J \longrightarrow k \longrightarrow 0.$$

Hence [k] + [k] = [R/J] = [k].

This implies [R/J] = 0.

<u>Case 2</u>. $\dim_k J < \infty$. Then we have the exact sequence $0 \longrightarrow J \longrightarrow R \longrightarrow R/J \longrightarrow 0$. But $[J] = [\sum_1^n k] = 0$ by the above and hence [R/J] = [R]. We make M into an R module by $e_i a = c_i$, $e_i c_j = 0$, and $e_i b = 0$ if i is even and $e_i b = c_i$ if i is odd. Then we have the exact sequence

$$0 \longrightarrow R \longrightarrow M \longrightarrow k \longrightarrow 0 \qquad \text{by sending 1 to a.}$$

We also have the sequence

$$0 \longrightarrow R/(e_2, e_4, \ldots, e_{2m}, \ldots) \longrightarrow M \longrightarrow R/(e_1, e_3, \ldots, e_{2n+1}, \ldots) \longrightarrow 0$$

by sending 1 to b .

Hence $[R] = [M] - [k] = 0$ by case 1.

Therefore, K_0 is 0 as claimed.

Remark. In the above example if we restrict to the finitely presented modules then K_0 is isomorphic to $Z \oplus Z$ with generators $[R]$ and $[R/(e_1)]$. If we try restricting to the coherent R modules, we also fail since R has only 0 as a coherent module. Therefore, K_0 of this is 0.

Now we want to discuss under what conditions a map of rings induces a map on G_0. Let $h: R \longrightarrow R'$ and M an R module then $R' \otimes_R M$ is an R' module. If R' is a flat right R module then tensoring with R' preserves exact sequences and, hence, relations in $G_0(R)$. The problem is to check if noetherian modules go to noetherian modules.

1) If R' is noetherian and M a finitely generated R module, then $R' \otimes_R M$ is a finitely generated R' module and hence noetherian. Thus we get an induced map in G_0.

2) If R and R' are commutative and R' is finitely generated as an R algebra or if R' is generated as a ring by the image of R and a finite number of central elements, then we get an induced map via $R' \otimes_R$. The proof of this follows from the remainder of this chapter.

We can also use h to give R' modules an R structure. Then we can map an R' module M to M as an R module. M will be a noetherian R module if

1) h is onto or

2) R' is a noetherian R module.

Let $\underline{A}$ be an abelian category. We construct a new category $\underline{\underline{A}[x]}$. The objects are pairs (A, f) where $A \in \underline{A}$ and $f \in \mathrm{End}(A)$. A morphism H from (A, f) to (B, g) is an element, H, of $\mathrm{Hom}_{\underline{A}}(A, B)$ such that

$$\begin{array}{ccc} A & \xrightarrow{H} & B \\ {\scriptstyle f}\downarrow & & \downarrow{\scriptstyle g} \\ A & \xrightarrow{H} & B \end{array} \qquad \text{commutes.}$$

This is an abelian category.

Definition. Let $\underline{A}$ be a countably right complete abelian category and $A \in \underline{A}$. Then $\underline{A[x]} = (X, f)$ where $X = \coprod_{0}^{\infty} A$ and

$f(x_0, x_1, x_2, \ldots) = (0, x_0, x_1, \ldots)$.

This gives a functor $\underline{A} \longrightarrow \underline{\underline{A}[x]}$.

The following generalization of Hilbert's basis theorem was stated by Evens for the case of modules.

Theorem 3.5. (Hilbert's basis theorem). Let $\underline{A}$ be a countably right complete abelian category in which countable directed $\varinjlim$ are exact. Then if A is noetherian in $\underline{A}$, A[x] is noetherian in $\underline{\underline{A}[x]}$.

Proof. Let (X, f) be as in the definition above and $X^n = \sum_0^n A$. Let B be a subobject of X which is stable under f, $B^{(n)} = X^n \cap B$, and $\overline{B^{(n)}}$ be the coker of the injection $B^{(n-1)} \longrightarrow B^{(n)}$ identified with a subobject of $A = \mathrm{ckr}[X^{n-1} \longrightarrow X^n]$. Then $f(B^{(n)}) \subseteq B^{(n+1)}$ and $\overline{B}^{(n)} \subseteq \overline{B}^{(n+1)}$ since B is stable under f.

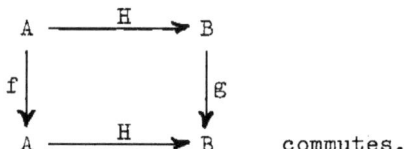

Let $B_1 \leq B_2 \leq \ldots .$ be a sequence of subobjects of $A[x]$. The $\overline{B^{(n)}}$ form a nonempty collection of subobjects of A. Let $\overline{B_{j_0}^{(n_0)}}$ be a maximal one. Thus, if $n \geq n_0$ and $j \geq j_0$, then $\overline{B_j^n} = \overline{B_{j_0}^{n_0}}$. Since there are only a finite number of $n \leq n_0$ we can pick a J so that if $j \geq J$ then $\overline{B}_j^{(n)} = \overline{B}_J^{(n)}$ for all n.

Then by induction on n and the five lemma applied to the diagram

$$\begin{array}{ccccccccc} 0 & \longrightarrow & B_J^{(n-1)} & \longrightarrow & B_J^{(n)} & \longrightarrow & \overline{B_J^{(n)}} & \longrightarrow & 0 \\ & & \downarrow & & \downarrow & & \downarrow & & \\ 0 & \longrightarrow & B_j^{(n-1)} & \longrightarrow & B_j^{(n)} & \longrightarrow & \overline{B_j^{(n)}} & \longrightarrow & 0 \end{array}$$

we obtain that $B_j^{(n)} = B_J^{(n)}$ for all n and for all $j \geq J$.

We claim $B_j = B_J$ for all $j \geq J$. $B^{(n)} \subset B$. Hence we get a map $\varinjlim B^{(n)} \longrightarrow B$. $0 \longrightarrow B^{(n)} \longrightarrow B \longrightarrow X/X^{(n)}$ is exact and directed $\varinjlim$ is exact hence

$$0 \longrightarrow \varinjlim B^{(n)} \longrightarrow B \longrightarrow \varinjlim X/X^{(n)} \qquad \text{is exact.}$$

Therefore $B = \varinjlim B^{(n)}$ and $B_j = \varinjlim B_j^{(n)}$.

Hence $B_j = B_J$ for $j \geq J$ since they are equal at each step.

Done.

<u>Remark</u>. This corresponds to one of the usual proofs of the Hilbert basis theorem where f is multiplication by x.

Now we generalize this slightly. Let Γ be a semigroup. We define $\underline{A}[\Gamma]$ to be the category whose objects are pairs (A, F) where $A \in \underline{A}$ and $F: \Gamma \longrightarrow \mathrm{End}(A)$ is multiplicative and preserves unit. A morphism of (A, F) to (B, G) is an element $f \in \mathrm{Hom}(A, B)$ such that

$$\begin{array}{ccc} A & \xrightarrow{f} & B \\ {\scriptstyle F(g)}\downarrow & & \downarrow{\scriptstyle G(g)} \\ A & \xrightarrow{f} & B \end{array}$$

commutes for all $g \in \Gamma$.

If Γ is a free semigroup on one generator then $\underline{A}[\Gamma] = \underline{A[x]}$. If Γ is a free commutative semi group on $x_1, \ldots, x_n$ then $\underline{A[\Gamma]} = \underline{A[x_1, \ldots, x_n]} = A[x_1, \ldots, x_{n-1}][x_n]$. In this case an object of $\underline{A}[\Gamma]$ is an object of $\underline{A}$ with n commuting endomorphisms.

Next we construct a functor from $\underline{A}$ to $\underline{A}[\Gamma]$ when Γ is finitely generated and commutative. This will be a left adjoint of the forgetful functor $\underline{A}[\Gamma] \longrightarrow \underline{A}$. For $A \in \underline{A}$ let $A[\Gamma] = \sum_{g \in \Gamma} A$. For each $g \in \Gamma$ we have the canonical injection $i_g: A \longrightarrow A[\Gamma]$. Let $F: \Gamma \longrightarrow \mathrm{End}(A[\Gamma])$ be given by $F(g)$ is the map making the following commute:

$$\begin{array}{ccc} A[\Gamma] & \xrightarrow{F(g)} & A[\Gamma] \\ {\scriptstyle i_{g'}}\nwarrow & & \nearrow{\scriptstyle i_{gg'}} \\ & A & \end{array}$$

for all $g' \in \Gamma$.

<u>Remarks</u>. More intuitively we index by Γ and use the group action to push the indexing around. If $\Gamma \longrightarrow \Gamma'$ is an onto homomorphism of semi groups then $\underline{A}[\Gamma']$ is a full subcategory of $\underline{A}[\Gamma]$ and the obvious map of $A[\Gamma]$ to $A[\Gamma']$ is an epi in $\underline{A}[\Gamma]$ for all $A \in \underline{A}$ if directed $\varinjlim$ is exact in $\underline{A}$.

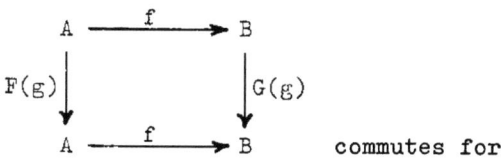

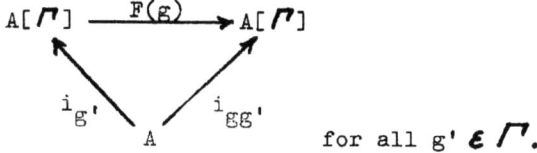

Theorem 3.6. Let $\underline{A}$ be a countably right complete abelian category whi has exact countable directed $\varinjlim$ and let G be a finitely generated commutative semigroup. If $A \in \underline{A}$ is noetherian, then A[G] is noetherian in $\underline{A}$[G].

Proof. Induction on n and the previous theorem establishes the result when G is free. The general case follows by expressing G as the image of a finitely generated free commutative semigroup and using the above remarks. Done.

Remark. This in particular takes care of the case of G infinite cyclic. We will return to this when we define $K_1(\underline{A})$.

Corollary 3.7. Let $h: R \longrightarrow R'$ be a ring homomorphism. Suppose R is generated by h(R) and a finite number of central elements. Then M a noetherian R module implies $R' \otimes_R M$ is a noetherian R' module.

Proof. R' is a quotient of $R[x_1, \ldots, x_n] = R''$. Let $\underline{A}$ be the category of R modules and G the free commutative semigroup on $\{x_1, \ldots, x_n\}$. Then $R'' \otimes_R M$ is isomorphic to M[G]. The theorem implies that $R'' \otimes_R M$ is noetherian. But we have an epi $R'' \longrightarrow R'$. Therefore we get an epi $R'' \otimes_R M \longrightarrow R' \otimes_R M$. Hence $R' \otimes_R M$ is noetherian as an R' module. Done.

Remark. Therefore if R and R' are as above and R' is a flat R module, $R' \otimes_R$ ___ induces a homomorphism from $G_0(R)$ to $G_0(R')$ as predicted.

Chapter 4. The Connection Between $K_0(R)$ and $G_0(R)$.

Lemma 4.1. (Schanuel). Let $\underline{A}$ be an abelian category. Let $0 \longrightarrow B \xrightarrow{i} P \xrightarrow{j} A \longrightarrow 0$ and $0 \longrightarrow B' \xrightarrow{i'} P' \xrightarrow{j'} A' \longrightarrow 0$ be exact with A isomorphic to A' and P and P' projective. Then $B \oplus P'$ is isomorphic to $B' \oplus P$.

<u>Proof</u>. We form the map of exact sequences

$$\begin{array}{ccccccccc} 0 & \longrightarrow & B & \xrightarrow{i} & P & \xrightarrow{j} & A & \longrightarrow & 0 \\ & & \downarrow g & & \downarrow f & & \downarrow h & & \\ 0 & \longrightarrow & B' & \xrightarrow{i'} & P' & \xrightarrow{j'} & A' & \longrightarrow & 0 \end{array}$$

where

h is an isomorphism, f exists since P is projective, and g exists since $j'fi = 0$ and B' is the kernel of j'. Then we form the sequence

$$0 \longrightarrow B \xrightarrow{(g,i)} B' \oplus P \xrightarrow{(i',-f)} P' \longrightarrow 0$$

which we prove is exact by applying the embedding theorem. Then P' projective implies the sequence splits and that $B \oplus P'$ is isomorphic to $B' \oplus P$. Done.

<u>Corollary 4.2</u>. Let $0 \longrightarrow B_n \longrightarrow P_{n-1} \longrightarrow \dots \longrightarrow P_0 \longrightarrow A \longrightarrow 0$ and $0 \longrightarrow B'_n \longrightarrow P'_{n-1} \longrightarrow \dots \longrightarrow P'_0 \longrightarrow A' \longrightarrow 0$ be exact in $\underline{A}$ with P_i and P'_i projective and A isomorphic to A'. Then $B_n \oplus P'_{n-1} \oplus P_{n-2} \oplus \dots$ is isomorphic to $B'_n \oplus P_{n-1} \oplus P'_{n-2} \oplus \dots$. and B_n is projective if and only if B'_n is.

<u>Proof</u>. The first follows by a straightforward induction on n. The second is obvious. Done.

<u>Definition</u>. Let $A \in \underline{A}$, an abelian category. The <u>projective dimension of A</u> (pd(A)) is the smallest n such that there exists an exact sequence $0 \longrightarrow P_n \longrightarrow P_{n-1} \longrightarrow \dots \longrightarrow P_0 \longrightarrow A \longrightarrow 0$ with all the P_i projective. If the projective dimension of A is undefined we write $pd(A) < \infty$.

The following corollary shows that any projective resolution of A may be used to calculate pd(A).

Corollary 4.3. Let $0 \to B_n \to P_{n-1} \to \dots P_0 \to A \to 0$ be exact with the P_i all projective. Then $pd(A) \leq n$ if and only if B_n is projective.

Proof. $\to$ by definition.

$\Leftarrow$ If $pd(A) \leq n$ then there is a resolution with P_i all projective and B_n' projective. Corollary 4.2 proves that B_n is also projective. Done.

Theorem 4.4. Let $\underline{A}$ be an abelian category and $\underline{P}$ be the full subcategory of projective objects. If $pdA < \infty$ for every $A \in \underline{A}$, then the natural map $I\colon K_0(\underline{P}) \to K_0(\underline{A})$ is an iso.

Proof. We construct the inverse $f\colon K_0(\underline{A}) \to K_0(\underline{P})$. If $A \in \underline{A}$, let $0 \to P_n \to P_{n-1} \to \dots P_0 \to A \to 0$ be a projective resolution of A. Let $f(A) = \sum_0^n (-1)^i [P_i]$.

First we must show that f is well-defined.

Let $0 \to P_n' \to P_{n-1}' \to \dots \to P_0' \to A \to 0$ be another projective resolution of A of the same length. Schanuel's lemma says that $P_n \oplus P_{n-1}' \oplus \dots$ is isomorphic to $P_n' \oplus P_{n-1} \oplus \dots$. That is $\sum_{i \text{ odd}} P_i \oplus \sum_{j \text{ even}} P_j'$ is isomorphic to $\sum_{i \text{ even}} P_i \oplus \sum_{j \text{ odd}} P_j'$. Therefore $\sum (-1)^i [P_i]$ is isomorphic to $\sum (-1)^j [P_j']$ and f(A) is well defined.

Next we show that f satisfies all the relations. Let $0 \to A \xrightarrow{i} A \xrightarrow{j} A'' \to 0$ be an exact sequence and $0 \to P_n' \to \dots \to P_0' \xrightarrow{p_0'} A' \to 0$ and $0 \to P_n'' \to P_{n-1}'' \to \dots \to P_0'' \xrightarrow{p_0''} A'' \to 0$ be projective resolution for A' and A''. Then $P_0' \oplus P_0''$ is projective. p_0'' is an epi. Hence, there exists

$h_0 \colon P_0'' \longrightarrow A$ such that

$$\begin{array}{ccccc} & & P_0'' & & \\ & \overset{h_0}{\swarrow} & \downarrow{\scriptstyle p_0''} & & \\ A & \xrightarrow{j} & A'' & \longrightarrow & 0 \end{array} \qquad \text{commutes.}$$

Then the pair $(ip_0', h_0)\colon P_0' \oplus P_0'' \longrightarrow A$ is an epi by the 5 lemma. Then look at the sequence of kernels $0 \longrightarrow B_0' \longrightarrow B_0 \longrightarrow B_0'' \longrightarrow 0$

$$\begin{array}{ccccccccc} & & 0 & & 0 & & 0 & & \\ & & \downarrow & & \downarrow & & \downarrow & & \\ 0 & \longrightarrow & B_0' & \longrightarrow & B_0 & \longrightarrow & B_0'' & \longrightarrow & 0 \\ & & \downarrow & & \downarrow & & \downarrow & & \\ 0 & \longrightarrow & P_0' & \longrightarrow & P_0' \oplus P_0'' & \longrightarrow & P_0'' & \longrightarrow & 0 \\ & & \downarrow & & \downarrow & & \downarrow & & \\ 0 & \longrightarrow & A' & \longrightarrow & A & \longrightarrow & A'' & \longrightarrow & 0 \end{array}$$

of the maps to the A's. This sequence is exact by the 9 lemma. Therefore we can start the process again. Finally, we get a projective resolution for A of the form

$$0 \longrightarrow P_n' \oplus P_n'' \longrightarrow \ldots \longrightarrow P_0' \oplus P_0'' \longrightarrow A \longrightarrow 0.$$

Then, $f(A) = \sum(-1)^i[P_i' \oplus P_i''] = \sum(-1)^i([P_j'] + [P_i'']) = f(A') + f(A'')$. Hence, f is additive and f is well defined.

Let $P \in \underline{P}$. Then $0 \longrightarrow P \longrightarrow P \longrightarrow 0$ is a projective resolution of P and $fI([P]) = [P]$.

Let $A \in \underline{A}$ and let $0 \longrightarrow P_n \longrightarrow \ldots \longrightarrow P_1 \longrightarrow P_0 \longrightarrow A \longrightarrow 0$ be a projective resolution for A. $If([A]) = \sum(-1)^i[P_i]$. Let B_0 be the kernel of $P_0 \longrightarrow A$. Then $0 \longrightarrow P_n \longrightarrow \ldots \longrightarrow P_1 \longrightarrow B_0 \longrightarrow 0$ is exact and induction on n proves that $[B_0] = \sum_{i=1}^{n}(-1)[P_i]$. But

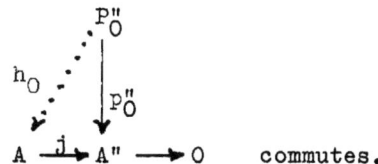

$0 \longrightarrow B_0 \longrightarrow P_0 \longrightarrow A \longrightarrow 0$ exact implies $[P_0] = [B_0] + [A]$. Therefore, $[A] = (\sum_{i=1}^{n}(-1)^i[P_i]) + [P_0] = If([A])$.

Hence I and f are inverse isomorphisms. Done.

If $\underline{A}$ is an abelian category, $\underline{P}$ the full subcategory of projective objects, and $\underline{H}$ the full subcategory of all $A \in \underline{A}$ with $pdA < \infty$ ($\underline{H}$ can fail to an abelian category), then the above proof yields the following:

Corollary 4.5. The natural map $I: K_0(\underline{P}) \longrightarrow K_0(\underline{H})$ is an iso. If $\underline{B}$ is a full subcategory $\underline{P} \subset \underline{B} \subset \underline{H}$, then the natural map $K_0(\underline{P}) \longrightarrow K_0(\underline{B})$ is a split mono. It is onto if for every exact sequence $0 \longrightarrow C \longrightarrow P \longrightarrow B \longrightarrow 0$. $P \in \underline{P}$ and $B \in \underline{B}$ implies $C \in \underline{B}$.

Definition. A ring R is left regular if it is left noetherian and every finitely generated left R module, M, has $pdM < \infty$.

Corollary 4.6. If R is left regular, the natural map $K_0(R) \longrightarrow G_0(R)$ is an iso.

Proof. Let $\underline{A}$ be the category of finitely generated left R modules and $\underline{P}$ be the category of finitely generated projective R modules. Then objects of $\underline{A}$ have finite resolutions by objects of $\underline{P}$. Hence, theorem 4.4 applies. Done.

Definition. The global dimension of a ring R is $\leq n$ if for all R modules A, $pdA \leq n$.

Corollary 4.7. If R is left noetherian and has finite global dimension, then $K_0(R)$ is isomorphic to $G_0(R)$.

Now we wish to prove that if R is left regular, then R[x] is. Theorem 3.5. proves that R[x] is left noetherian.

<u>Definition</u>. Let $A \in \underline{A}$, an abelian category. Then A is a <u>projectively resolvable object</u> if A has a (possibly infinite) projective resolution.

<u>Proposition 4.8</u>. Let $\underline{A}$ be an abelian category and A', A, and $A'' \in \underline{A}$ with $0 \longrightarrow A' \longrightarrow A \longrightarrow A'' \longrightarrow 0$ exact. Then

1) If any two are projectively resolvable, then so is the third,

2) If any two have finite projective dimension, so does the third, and

3) a) pdA' and pdA" $\leq$ n implies pdA $\leq$ n,

   b) pdA and pdA" $\leq$ n implies pdA' $\leq$ n, and

   c) pdA' and pdA $\leq$ n implies pdA" $\leq$ n + 1.

<u>Proof</u>. We omit the proof of 1) since it doesn't arise in the case of modules. 3) $\Longrightarrow$ 2). Hence we prove 3) assuming all three objects have a finite projective resolution. Then we obtain a commutative diagram:

$$\begin{array}{ccccccccc}
 & & 0 & & 0 & & 0 & & \\
 & & \downarrow & & \downarrow & & \downarrow & & \\
0 & \longrightarrow & B'_n & \longrightarrow & B_n & \longrightarrow & B''_n & \longrightarrow & 0 \\
 & & \downarrow & & \downarrow & & \downarrow & & \\
0 & \longrightarrow & P'_n & \longrightarrow & P_n & \longrightarrow & P''_n & \longrightarrow & 0 \\
 & & \downarrow & & \downarrow & & \downarrow & & \\
 & & \vdots & & \vdots & & \vdots & & \\
 & & \downarrow & & \downarrow & & \downarrow & & \\
0 & \longrightarrow & P'_0 & \longrightarrow & P_0 & \longrightarrow & P''_0 & \longrightarrow & 0 \\
 & & \downarrow & & \downarrow & & \downarrow & & \\
0 & \longrightarrow & A' & \longrightarrow & A & \longrightarrow & A'' & \longrightarrow & 0 \\
 & & \downarrow & & \downarrow & & \downarrow & & \\
 & & 0 & & 0 & & 0 & &
\end{array}$$

where all rows and columns are exact and where all P_i, P'_i, and P''_i are projective.

3a) If pdA' and pdA" $\leq$ n then B'_n and B''_n are projective. Hence

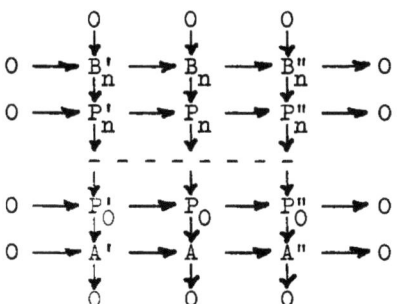

$0 \longrightarrow B_n' \longrightarrow B_n \longrightarrow B_n'' \longrightarrow 0$ splits and B_n is projective. Therefore, $\mathrm{pd}A \leqslant n$.

3b) if pdA and pdA" $\leqslant n$, then B_n and B_n'' are projective. Hence $0 \longrightarrow B_n' \longrightarrow B_n \longrightarrow B_n'' \longrightarrow 0$ splits and B_n' is projective. Therefore, $\mathrm{pd}A \leqslant n$.

3c) if pdA' and pdA $\leqslant n$, then B_n and B_n' are projective. The composite $B_n \longrightarrow B_n'' \longrightarrow P_n''$ gives an exact sequence

$$0 \longrightarrow B_n' \longrightarrow B_n \longrightarrow P_{n-1}'' \longrightarrow \cdots \longrightarrow A'' \longrightarrow 0$$

which proves pdA" $\leqslant n + 1$. Done.

Theorem 4.9. Let $\underline{A}$ be an abelian category with countable directed $\varinjlim$ exact and countably right complete. If $(A, \alpha) \in \underline{A[x]}$. Then $\mathrm{pd}(A, \alpha) \leqslant 1 + \mathrm{pd}A$.

Proof. For $A \in \underline{A}$ set $A[x] = F(A) = (\sum_0^\infty A, f)$ where f is the shift map $f(x_0, x_1, \ldots.) = (0, x_0, x_1, \ldots.)$. Then F can be made into a functor in the obvious way. The functor $G: \underline{A[x]} \longrightarrow \underline{A}$ which on objects is $G((A, \alpha)) = A$. F and G are adjoints. That is

$$\mathrm{Hom}_{\underline{A[x]}}(A[x], (B, b)) \cong \mathrm{Hom}_{\underline{A}}(A, B).$$

Therefore F is exact and preserves projectives. To finish the theorem we need the following:

Lemma 4.10. Let $(A, f) \in \underline{A[x]}$. Then there exists an exact sequence

$$0 \longrightarrow A[x] \longrightarrow A[x] \longrightarrow (A, f) \longrightarrow 0 \text{ in } \underline{A[x]}.$$

Proof. Assuming $\underline{A}$ has elements we construct the maps explicitly:

$$0 \longrightarrow A[x] \xrightarrow{i} A[x] \xrightarrow{j} (A,\ f) \longrightarrow 0\ .$$

where $i(a_0,\ a,\ \ldots) = (f(a_0),\ f(g) - a_0,\ f(a_2) - a_1,\ \ldots)$ and

$$j(a_0,\ a_1,\ \ldots) = a_0 + f(a_1) + f^2(a_2) + \ldots + f^n(a_n) + \ldots\ .$$

By inspection this sequence is exact. It is easy to check that i and j are admissible maps. The general result applies in any abelian category by using the imbedding theorem and exact limits.

The maps can be defined directly by letting $j\colon \coprod_0^{\infty} A \longrightarrow A$ be f^n on the n-th factor and defining i to be

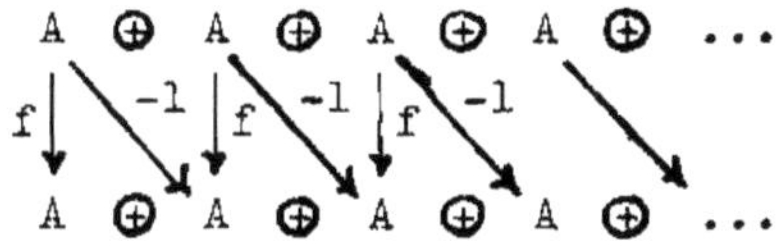

Done with lemma 4.10.

Since F preserves projectives, $pd_{\underline{A}[x]} A[x] \leqslant pd_{\underline{A}} A$.

$pd_{\underline{A}[x]} A[x] \leqslant n$ implies $pd_{\underline{A}[x]}(A,\ \alpha) \leqslant n + 1$ by lemma 4.10 and proposition 4.8. Done with theorem 4.9.

<u>Lemma 4.11</u>. Let $\underline{A}$ be an abelian category as in theorem 4.9. If we have a sequence

$$A_0 \xrightarrow{f_0} A_1 \xrightarrow{f_1} A_2 \longrightarrow \ldots \longrightarrow A_n \xrightarrow{f_n} A_{n+1} \longrightarrow \ldots$$

with $\varinjlim A_n = A$, then there is an exact sequence

$$* \qquad 0 \longrightarrow \coprod_0^{\infty} A_n \xrightarrow{i} \coprod_0^{\infty} A_n \xrightarrow{j} A \longrightarrow 0\ .$$

<u>Proof</u>. Assume as before that $\underline{A}$ has elements

$$i(a_0,\ A_1,\ \ldots.) = (a_0,\ a_1 - f_0(a_0),\ a_2 - f_1(a_1),\ \ldots.)$$

$$j(a_0,\ a_1,\ \ldots.) = g(a_0) + g_1(a_1) + \ldots. + g_n(a_n)$$

where $g_n\colon A_n \longrightarrow A$ is the natural map. Then * is exact. The maps are defined as in lemma 4.10. Done.

Theorem 4.12. Under the assumption of lemma 4.11 $pdA_i \leqslant k$ implies $pdA \leqslant k + 1$.

Proof. For each A_n there is a projective resolution

$$0 \longrightarrow P_k^{(n)} \longrightarrow \ldots. \longrightarrow P_0^{(n)} \longrightarrow A_n \longrightarrow 0.$$

But $\mathrm{Hom}(\coprod P_i,\ \_\_) = \prod \mathrm{Hom}(P_i,\ \_\_)$. Hence $\coprod P_i$ is projective if each P_i is. But then the sequence

$$0 \longrightarrow \coprod P_k^{(n)} \longrightarrow \ldots \longrightarrow \coprod P_0^{(n)} \longrightarrow \coprod A_n \longrightarrow 0 \quad \text{is exact.}$$

Hence $pd\coprod A_n \leqslant k$. Lemma 4.11 and proposition 4.8 finish the proof.

Done.

Theorem 4.13. Let R be a left regular ring. Then R[x] is also.

Proof. Let M be a finitely generated R[x] module with $m_1, \ldots, m_n$ a set of generators. Let N be the R submodule of M generated by the m_i. $M = \sum_{i=0} x^i N$. Let $M_n = \sum_{i=0}^{n} x^i N$. Then we have an ascending sequence of R modules $M_0 \subset M_1 \subset \ldots.. .$ If $\hat{x_i}$ denotes the map obtained by multiplication by x^i the following diagram commutes.

$$\begin{array}{ccccccc}
N & \xrightarrow{\hat{x^n}} & M_n & \longrightarrow & M_n/M_{n-1} & \longrightarrow & 0 \\
{\scriptstyle 1_N}\downarrow & & \downarrow{\scriptstyle \hat{x}} & & \downarrow{\scriptstyle \hat{x}} & & \\
N & \xrightarrow[\hat{x^{n+1}}]{} & M_{n+1} & \longrightarrow & M_{n+1}/M_n & \longrightarrow & 0
\end{array} \quad .$$

If K_n is the kernel of $N \longrightarrow M_n/M_{n-1}$, then the ascending sequence $K_0 \subset K_1 \subset \dots \subset K_n \subset$ must terminate. Therefore there exists n_0 such that $n \geq n_0$ implies M_{n+1}/M_n is isomorphic to M_{n_0}/M_{n_0-1}.

Let $r = \max[pd_R(M_{n_0}, pd_R(M_{n_0}/M_{n_0-1})]$. If $n \geq n_0$, an easy inductive argument shows that $pd_R M_n \leq r$.

But $M = \underset{\longrightarrow}{\lim} M_n$ implies $pd_R M \leq r + 1$ by theorem 4.12. Theorem 4.9 implies that $pd_{R[x]} M \leq r + 2$. Hence $R[x]$ is left regular. Done.

Chapter 5. Localization and Relation between $G_0(R)$ and $G_0(R_S)$.

In this chapter R is a ring and S is a multiplicatively closed subset of the center of R which contains 1.

If M is an R module we define a new R module M_S which has elements equivalence classes of pairs (m, s) where $m \in M$ and $s \in S$ and $(m, s) \sim (m', s')$ if there exists $t \in S$ such that $t(s'm - sm') = 0$.

If m/s is the class of (m, s) then we define

$$\frac{m}{s} + \frac{m'}{s'} = \frac{s'm + sm'}{ss'},$$

$$r\,\frac{m}{s} = \frac{rm}{s} \qquad \text{for all } r \in R.$$

If M = R then R_S is a ring relative to the multiplication

$$\frac{r}{s}\,\frac{r'}{s'} = \frac{rr'}{ss'} \quad .$$

M_S is an R_S module with multiplication $\frac{r}{s}\,\frac{m}{t} = \frac{rm}{st}$.

The map $f\colon M \longrightarrow M_S$ given by $f(m) = \frac{m}{1}$ for $m \in M$ is an R map. If M = R it is a ring homomorphism. $\mathrm{Ker}(f) = \{m \in M \mid \text{there exists } s \in S \text{ with } sm = 0\}$.

f is universal for R module maps of M to R modules, N, on which S acts by isos. $R \longrightarrow R_S$ is universal for m ring homomorphism sending S to units.

It is easy to see that for any R module M, the R-map $M \longrightarrow R_S \otimes_R M$ given by m goes to $1 \otimes m$ is universal for R maps to modules on which S acts by isos. Hence M_S is isomorphic to $R_S \otimes_R M$. This makes $M \rightsquigarrow M_S$ into a functor.

<u>Proposition 5.1</u>. The functor $M \rightsquigarrow M_S$ is exact (that is, R_S is a flat R module).

<u>Proof</u>. Let $0 \longrightarrow M' \xrightarrow{i} M \xrightarrow{j} M'' \longrightarrow 0$ be an exact sequence of R modules. Since tensor is right exact we only need to check that the map $M'_S \xrightarrow{i_S} M_S$ is a mono. Suppose $i_S(\frac{m'}{s}) = 0$. Then there exists $t \in S$ such that $ti(m') = 0$. Hence $tm' = 0$. Therefore in m'_S, $\frac{m'}{s} = 0$. Done.

In the canonical map $f\colon M \longrightarrow M_S$, suppose N is an R_S submodule of M_S. Then $(f^{-1}(N))_S = N$. For if $x \in (f^{-1}(N))_S$, $x = n/s$ where $n \in f^{-1}(N)$. Hence $n/1 \in N$ and $n/s \in N$. Thus $(f^{-1}(N))_S \subset N$. But if $\frac{n}{s} \in N$, then $\frac{sn}{s} = \frac{n}{1} \in N$. Therefore $n \in f^{-1}(N)$ and $\frac{n}{s} \in (f^{-1}(N))_S$.

<u>Corollary 5.2</u>. If M is a noetherian R module, then M_S is a noetherian R_S module. If R is a noetherian ring, then R_S is also.

<u>Proposition 5.3</u>. If R is left regular, then R_S is also.

<u>Lemma</u>. Let M be a finitely generated R_S module. Then M is isomorphic to N_S where N is a finitely generated R module.

<u>Proof</u>. Let $m_1, \ldots, m_n$ generate M over R_S. Let N be the R submodule of M generated by $m_1, \ldots, m_n$.

<u>Proof of 5.3</u>. N has a finite resolution by projective R modules:

$$0 \to P_n \to \dots \to P_0 \to N \to 0 \quad .$$

But R_S is a flat R module. Therefore,

$$0 \to P_{n_S} \to \dots . \to P_{0_S} \to N_S \to 0 \quad \text{is exact}$$

and all P_{i_S} are projective. Hence $pd_{R_S} M \leq pd_R N < \infty$. Done.

Corollary 5.4. If R is left regular, then $R[X, X^{-1}]$ is also.

Proof. R is left regular implies R[x] is by theorem 4.13. R[x] left regular implies $R[X, X^{-1}] = R[x]_S$ where $S = \{1, x, x^2, \dots\}$ is by proposition 5.3. Done.

Proposition 5.5. Let R be a commutative ring then the map $p \rightsquigarrow p_S$ gives a 1-1 correspondence between the prime ideals p of R such that $p \cap S = \emptyset$ and all the primes of R_S. The correspondence is lattice preserving, and the inverse map is given by $P \to f^{-1}(P)$ where f is the natural map $R \to R_S$.

Proof. If p is a prime ideal of R, we want that p_S is a prime ideal of RS. Let $\frac{r}{s}, \frac{r'}{s} \in R_S$ with $\frac{rr'}{ss'} \in p_S$. Then $\frac{rr'}{ss'} = \frac{p'}{t}$ where $p' \in p$ and $t \in S$. Then there exists $v \in S$ such that $v(trr'-ss'p') = 0$. Hence $vtrr' \in p$. But $p \cap S = \emptyset$ implies $rr' \in p$. Hence $r \in p$ or $r' \in p$ and $\frac{r}{s} \in p_S$ or $\frac{r'}{s'} \in p_S$. Therefore p_S is prime.

P is a prime ideal of R_S then $f^{-1}(P)$ is a prime ideal of R and $(f^{-1}(P))_S = P$. Let p be a prime ideal of R. Then $f^{-1}(p_S) \supset p$ is clear. Let $r \in f^{-1}(p_S)$. Then $f(r) = p'/s$ where $p' \in p$. Thus there exists $v \in S$ with $v(sr - p') = 0$. As before this implies that $r \in p$. Hence $f^{-1}(p_S) = P$. Therefore, the two maps are mutual inverses.

Done.

From now on R is a commutative ring with 1.

Proposition 5.6. If M and N are R modules, then $(M \otimes_R N)_S$ is isomorphic to $M_S \otimes_{R_S} N_S$.

Proof. Recall that if $f: R \longrightarrow R'$ is a homomorphism of commutative rings, then the tensor product associativity yields the isomorphism

$$R' \otimes_R (M \otimes_R N) \simeq (R' \otimes_R M) \otimes_{R'} (R' \otimes_R N).$$

Let $f: R \longrightarrow R_S$ be the map. Then $R_S \otimes M \simeq M_S$ yields the result.

Done.

Similarly from the natural map * $R' \otimes_R \mathrm{Hom}_R(M, N) \longrightarrow (R' \otimes_R M, R' \otimes_R$ we get a map

$$* \qquad \mathrm{Hom}_R(M, N)_S \longrightarrow \mathrm{Hom}_{R_S}(M_S, N_S)$$

given by

$$r' \otimes f \rightsquigarrow g \quad \text{where} \quad g(x \otimes m) = r' x \otimes f \subset m).$$

Definition. An R module M is finitely presented if there are finitely generated free R module F and F' and an exact sequence $F' \longrightarrow F \longrightarrow M \longrightarrow 0$.

Proposition 5.7. If M is a finitely presented R module and R' is a flat R module then the map * is an iso. In particular, it will be an iso for $R \longrightarrow R_S$ whenever M is finitely presented.

Proof. Let G( ) denote the contravariant additive functor $G(M) = R' \otimes_R \mathrm{Hom}_R(M, N)$ and let H( ) denote the functor $\mathrm{Hom}_{R'}(R' \otimes_R \_\_, R' \otimes_R N)$. Then * is an iso if

1) $M = R$ for then both sides are $R' \otimes_R N$.

2) M is a finitely generated free module. This is clear from 1) since both functors are additive.

3) M is finitely presented. Then we have $F' \longrightarrow F \longrightarrow M \longrightarrow 0$ exact we obtain the commutative diagram

$$\begin{array}{ccccccc} G(F') & \longleftarrow & G(P) & \longleftarrow & G(M) & \longleftarrow & 0 \\ \downarrow & & \downarrow & & \downarrow & & \\ H(P') & \longleftarrow & H(P) & \longleftarrow & H(M) & \longleftarrow & 0 . \end{array}$$

The rows are exact since R' is flat as an R module and the two left hand maps are isos. Therefore $G(M) \longrightarrow H(M)$ is an iso. Done.

Remark. 1) and 2) of the above show that the map is an iso whenever M is a finitely generated projective R module.

This proposition can be generalized slightly to the following:

Proposition 5.8. Let R be a commutative ring and A an R algebra. If $S \subset R$ is a nonempty multiplicative set and M and N are A modules with M finitely presented, then the natural map

$$\mathrm{Hom}_A(M, N)_S \longrightarrow \mathrm{Hom}_{A_S}(M_S, N_S) \quad \text{is an iso.}$$

Proof. Exactly as above.

Corollary 5.9. If A is an R algebra, M a finitely presented A module, N an A-module, and $f: M_S \longrightarrow N_S$ an A_S map, then there exists an A map $f: M \longrightarrow N$ and $s \in S$ so that $f = g/s$. That is the following diagram commutes

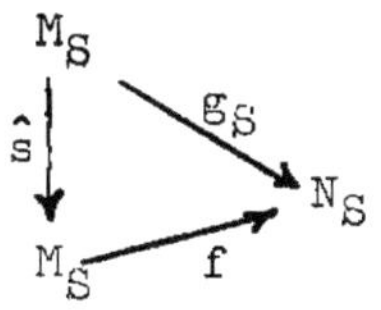

Proof. $f \in \mathrm{Hom}_{A_S}(M_S, N_S) = \mathrm{Hom}_A(M, N)_S$. Then there is $s \in S$ and $g \in \mathrm{Hom}_A(M, N)$ with $f = g/s$.

<u>Corollary 5.10</u>. If M is a finitely presented A_S module, there exists a finitely presented A module N such that N_S is isomorphic to M.

<u>Proof</u>. Let $F' \xrightarrow{f} F \longrightarrow M \longrightarrow 0$ be exact where F and F' are finitely generated free A_S modules. We let P and P' be free on the same number of generators. Then P_S is isomorphic to F and P'_S is isomorphic to F'. Then $f = g/s$ where $g\colon P' \longrightarrow P$ and $s \in S$. Then the following diagram shows that if N is the coker g, then N_S is isomorphic to M:

$$\begin{array}{ccccccc}
P' & \xrightarrow{g} & P_S & \longrightarrow & N_S & \longrightarrow & 0 \\
\Big\downarrow{\scriptstyle \hat{s}} & & \Big\downarrow{\scriptstyle 1} & & \Big\downarrow & & \Big\downarrow \\
P'_S & \longrightarrow & P_S & \longrightarrow & M & \longrightarrow & 0 \quad .
\end{array}$$

Done.

Let F be an exact covariant functor from $\underline{A}$ to $\underline{B}$, where $\underline{A}$ and $\underline{B}$ are abelian categories. Recall from Part I that $\underline{C} = \{C \in \underline{A} \mid F(C) = 0\}$ is a Serre subcategory and F factors $\underline{A} \xrightarrow{T} \underline{A}/\underline{C} \xrightarrow{G} B$ where T and $\underline{A}/\underline{C}$ were defined in Part I and G is exact.

<u>Theorem 5.11</u>. Suppose that in the above situation we also have:

1) For every $B \in \underline{B}$ there is an $A \in \underline{A}$ such that FA is isomorphic to B, and

2) If $f\colon FA \longrightarrow FA'$, then there exists a A'', $h\colon A'' \longrightarrow A$, and $g\colon A'' \longrightarrow A'$ such that the following diagram is commutative

$$\begin{array}{ccc}
 & FA'' & \\
{\scriptstyle Fh} \swarrow & & \searrow {\scriptstyle Fg} \\
FA & \xrightarrow[f]{} & FA'
\end{array}$$

and Fh is an iso.

Then $G\colon \underline{A}/\underline{C} \longrightarrow \underline{B}$ is an equivalence.

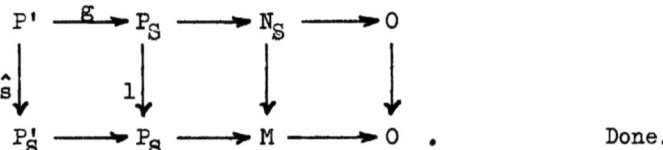

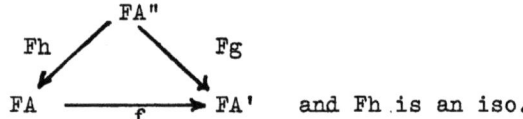

Proof. Since G is exact G is faithful if and only if GA = 0 implies A = 0. But GA = FA and FA = 0 means $A \in \underline{C}$. But $A \in \underline{C}$ means TA = 0 and that A is a zero object in $\underline{A}/\underline{C}$. Hence G is faithful.

Let f: GA ⟶ GA' then by 2) there exists g, h, A" such that the following commutes

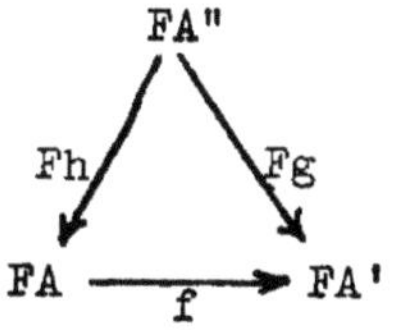

and such that Fh is an iso. Hence ker Fh and coker Fh = 0. Hence ker h and coker $h \in \underline{C}$. Therefore, Gh is an iso and $f = Gg(Gh)^{-1}$.

Hence h is full.

Therefore proposition I.1.18 implies G is an equivalence. Done.

Corollary 5.12. Let R be a noetherian ring, $\underline{A}$ be the category of all finitely generated R modules, $\underline{B}$ the category of all finitely generated R_S modules and $F(M) = M_S$. Then $\underline{C} = \{M \mid M$ is a finitely generated R module with $M_S = 0\}$ and $\underline{A}/\underline{C} \longrightarrow \underline{B}$ is an equivalence of categories.

Theorem 5.13. Let $\underline{A}$ and $\underline{B}$ be abelian categories, F: $\underline{A} \longrightarrow \underline{B}$ an exact functor, and $\underline{C} = \{A \in \underline{A} \mid FA = 0\}$. Assume that for if $B \in \underline{B}$ and, $A \in \underline{A}$, and f: B ⟶ FA, then there exists $A' \in A$ and g: A' ⟶ A, and h: FA' ⟶ B is an iso such that

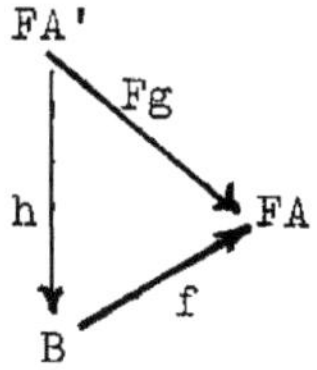

commutative.

Then

$$K_0(\underline{C}) \longrightarrow K_0(\underline{A}) \longrightarrow K_0(\underline{B}) \longrightarrow 0 \quad \text{is exact}$$

where the maps are induced from the inclusion and F respectively.

Corollary 5.14. If $\underline{C}$ is a Serre subcategory of an abelian category $\underline{A}$, then

$$K_0(\underline{C}) \longrightarrow K_0(\underline{A}) \longrightarrow K_0(\underline{A}/\underline{C}) \longrightarrow 0 \quad \text{is exact.}$$

Corollary 5.15. Let A be a noetherian R algebra where R is a commutative ring, $S \subset R$ be a multiplicative system, and $\underline{C}$ be the category of finitely generated A modules with $M_S = 0$. Then $K_0(\underline{C}) \longrightarrow G_0(A) \longrightarrow G_0(A_S) \longrightarrow 0$ is exact.

Remark. This does not prove that $K_0(\underline{C}) \longrightarrow K_0(A) \longrightarrow K_0(A_S) \longrightarrow 0$ is exact since the projectives don't form an abelian category. If A is left regular then $K_0(A)$ is naturally isomorphic to $G_0(A)$ and, hence,

$$K_0(\underline{C}) \longrightarrow K_0(A) \longrightarrow K_0(A_S) \longrightarrow 0 \quad \text{is exact.}$$

Proof of Theorem 5.13.

Let I be the image of $K_0(\underline{C})$ in $K_0(\underline{A})$. There is a map of $f\colon K_0(\underline{A})/I \longrightarrow K_0(\underline{B})$. We will prove the theorem by constructing a map $g\colon K_0(\underline{B}) \longrightarrow K_0(\underline{A})/I$ such that it is a left and right inverse of f.

Let $B \in \underline{B}$. Then there is an $A \in \underline{A}$ such that F(A) is isomorphic to B. (Apply the hypothesis to $B \longrightarrow F(0)$.) Let $g([B]) = [A]$ in $K_0(\underline{A})/I$. We must show that if $A' \in \underline{A}$ such that F(A') is isomorphic to B then $[A] = [A']$ in $K_0(\underline{A})/I$. The two isomorphisms give a map $B \longrightarrow F(A) \oplus F(A') = F(A \oplus A')$. Therefore, there is an $A'' \in \underline{A}$ and a map $h\colon A'' \longrightarrow A \oplus A'$ such that

$$\begin{array}{ccc} F(A'') & & \\ \downarrow \cong & \searrow^{F(h)} & \\ B & \longrightarrow & F(A \oplus A') \end{array} \qquad \text{commutes.}$$

Let $p_1 \colon A \oplus A' \longrightarrow A$ and $p_2 \colon A \oplus A' \longrightarrow A'$ be the projections. Then $F(p_1h)$ and $F(p_2h)$ are isos. Therefore, the ker and coker of p_1h are in $\underline{C}$. But

$$0 \longrightarrow \ker p_1h \longrightarrow A'' \longrightarrow A \longrightarrow \operatorname{coker} p_1h \longrightarrow 0 \quad \text{is exact.}$$

Therefore $[A] - [A''] = [\operatorname{coker} p_1h] - [\ker p_1h]$ in $K_0(\underline{A})$. Hence $[A] = [A'']$ in $K_0(\underline{A})/I$. Similarly $[A'] = [A'']$ in $K_0(\underline{A})/I$.

Next we check that if $0 \longrightarrow B' \longrightarrow B \xrightarrow{j} B'' \longrightarrow 0$ is exact. Then $g([B]) = g([B']) + g([B''])$. Pick $D \in \underline{A}$ such that B'' is isomorphic to $F(D)$. Then there is an $A \in \underline{A}$ and $h \colon A \longrightarrow D$ such that

$$\begin{array}{ccc} FA & \xrightarrow{F(h)} & FD \\ \Big| s & & \Big| s \\ B & \xrightarrow{j} & B'' \end{array} \qquad \text{commutes.}$$

h may not be an epi. Let A'' be the image of h. Then we have $0 \longrightarrow A'' \longrightarrow D \longrightarrow Q \longrightarrow 0$ exact and $F(Q) = 0$. Thus $F(A'')$ is isomorphic to $F(D)$ is isomorphic to B'' and $0 \longrightarrow A' \longrightarrow A \longrightarrow A'' \longrightarrow 0$ is exact where A' is the kernel of h. Applying the exact functor F we get

$$0 \longrightarrow F(A') \longrightarrow F(A) \longrightarrow F(A'') \longrightarrow 0 \quad \text{is exact.}$$

The five lemma show $F(A')$ is isomorphic to B'. Hence $g([B]) = [A] = [A'] + [A''] = g([B']) + g([B''])$.

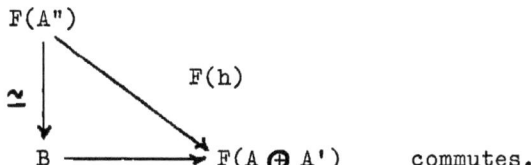

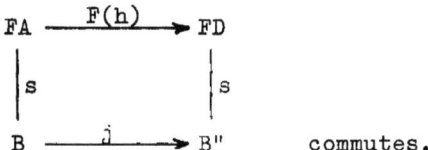

It is obvious that g is the right and left inverse of f. Done.

Let M be a finitely generated A module such that $M_S = 0$. Then for each generator m_i there is an $s_i \in S$ such that $s_i m_i = 0$. Let $s = \prod_{i=1}^{n} s_i$. Then $sM = 0$ and M is an A/sA module. Thus we get an exact sequence.

$$\sum_{s \in S} G_0(A/sA) \longrightarrow G_0(A) \longrightarrow G_0(A_S) \longrightarrow 0.$$

<u>Theorem 5.16</u>. If R is a commutative noetherian ring, S a multiplicatively closed subset of R and A a noetherian R-algebra then the sequence

$$\sum G_0(A/pA) \longrightarrow G_0(A) \longrightarrow G_0(A_S) \longrightarrow 0$$

is exact where the left hand sum is taken over all prime ideals p of R with $p \cap S \neq \emptyset$.

<u>Proof</u>. First we prove a lemma.

<u>Lemma 5.17</u>. If M is a finitely generated A module then there is a finite filtration, $M = M_n \supset M_{n-1} \supset \cdots \supset M_1 \supset M_0 = 0$ such that each M_i/M_{i-1} is a cyclic A module whose R annihilator is a prime ideal of R.

<u>Proof</u>. If $0 \longrightarrow M' \longrightarrow M \longrightarrow M'' \longrightarrow 0$ is exact and the lemma is true for M' and M" then it is true for M. Hence it is enough to look at the case where M is cyclic. Suppose the lemma is false for some cyclic module. Pick one whose R annihilator, $\underline{a}$, is maximal among all R annihilators of cyclic modules that makes the lemma false. $\underline{a}$ cannot be prime. Hence there are elements a and b of $R - \underline{a}$ such that $ab \in \underline{a}$. Consider the exact sequence

$$0 \longrightarrow aM \longrightarrow M \longrightarrow M/aM \longrightarrow 0.$$

The annihilator of aM is ($\underline{a}$, b) and the annihilator of M/aM is ($\underline{a}$, a). Hence the lemma holds for aM and M/aM. But then it holds for M. This is a contradiction. Done with lemma 5.17.

Back to theorem 5.16. The kernel of $G_0(A) \longrightarrow G_0(A_S)$ is generated by M such that $M_S = 0$. Then M is a finitely generated A/sA module for $s \in S$. Filter it as in the lemma.

$M = M_n \supset \dots \supset M_1 \supset M_0 = 0$. Then

$[M] = \sum [M_i/M_{i-1}]$ where

M_i/M_{i-1} is a finitely generated A module whose R annihilator is a prime ideal p_i. But sM = 0 implies $s(M_i/M_{i-1}) = 0$. Hence $s \in p_i$. Therefore $s \in p_i$. Done with theorem 5.16.

Theorem 5.18. Let R be a commutative noetherian ring. Then the maps $G_0(R) \longrightarrow G_0(R[X])$ and $G_0(R[X]) \longrightarrow G_0(R[X, X^{-1}])$ are isomorphisms.

Proof. First we map $G_0(R[X, X^{-1}]) \longrightarrow G_0(R)$ as follows: let M be a finitely generated $R[X, X^{-1}]$ module. Let K and Q be the ker and coker, respectively, of the map $M \longrightarrow M$ given by multiplication by x - 1. Then K and Q can be considered as finitely generated R modules since the action of x is trivial. The map to $G_0(R)$ sends [M] to [Q] - [K]. The following diagram and the 12-lemma show that this map is well defined.

$$
\begin{array}{ccccccccc}
 & & 0 & & 0 & & 0 & & \\
 & & \downarrow & & \downarrow & & \downarrow & & \\
0 & \longrightarrow & K' & \longrightarrow & K & \longrightarrow & K'' & \xrightarrow{d} & \\
 & & \downarrow & & \downarrow & & \downarrow & & \\
0 & \longrightarrow & M' & \longrightarrow & M & \longrightarrow & M'' & \longrightarrow & 0 \\
 & & \downarrow{\scriptstyle \widehat{x-1}} & & \downarrow{\scriptstyle \widehat{x-1}} & & \downarrow{\scriptstyle \widehat{x-1}} & & \\
0 & \longrightarrow & M' & \longrightarrow & M & \longrightarrow & M'' & \longrightarrow & 0 \\
 & & \downarrow & & \downarrow & & \downarrow & & \\
 & \xrightarrow{d} & Q' & \longrightarrow & Q & \longrightarrow & Q'' & \longrightarrow & 0 \\
 & & \downarrow & & \downarrow & & \downarrow & & \\
 & & 0 & & 0 & & 0 & &
\end{array}
$$

Since $M[X, X^{-1}] \xrightarrow{\widehat{x-1}} M[X, X^{-1}]$ has ker 0 and coker M the composite of the maps $G_0(R) \longrightarrow G_0(R[X, X^{-1}]) \longrightarrow G_0(R)$ is the identity on $G_0(R)$.

Since we already know that the map of $G_0(R[X]) \longrightarrow G_0(R[X, X^{-1}])$ is onto it will suffice to show that the map $G_0(R) \longrightarrow G_0(R[X])$ is onto. We will prove this by contradiction. Suppose it is false, then there is an ideal $\underline{a}$ of R such that $G_0(R/\underline{a}) \longrightarrow G_0(R/\underline{a}[X])$ is not onto and such that $\underline{a}$ is maximal with respect to this property. We replace R by $R/\underline{a}$. Hence we can assume the map is not onto for R but is for $R/\underline{c}$ for any non zero ideal $\underline{c}$. If R is not a domain, let S = R be a multiplicative system. Then we have the diagram:

$$
\begin{array}{ccccc}
\sum\limits_{\underline{p} \cap S \neq \emptyset} G_0(R/\underline{p}) & \longrightarrow & G_0(R) & \longrightarrow & 0 \\
\downarrow & & \downarrow & & \\
\sum\limits_{\underline{p} \cap S \neq \emptyset} G_0[(R/\underline{p})[X]] & \longrightarrow & G_0(R[X]) & \longrightarrow & 0
\end{array}
\quad .
$$

Since 0 is not a prime ideal, we have that the map on the left is onto. But the rows are exact, and the diagram commutes. Therefore, $G_0(R) \longrightarrow G_0(R[X])$ is onto.

If R is a domain, let $S = R - \{0\}$. Then we have the diagram

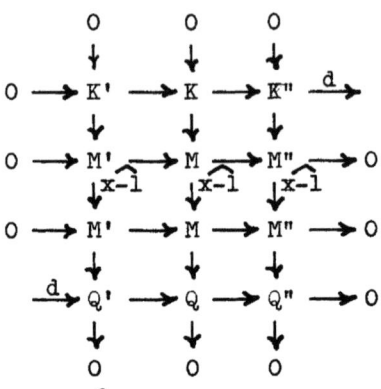

$$\begin{array}{ccccccc}
\sum_{\underline{p} \cap S \neq \emptyset} G_0(R/\underline{p}) & \longrightarrow & G_0(R) & \longrightarrow & G_0(K) & \longrightarrow & 0 \\
\downarrow & & \downarrow & & \downarrow & & \\
\sum_{\underline{p} \cap S \neq \emptyset} G_0((R/\underline{p})[X]) & \longrightarrow & G_0(R[X]) & \longrightarrow & G_0(K[X]) & \longrightarrow & 0
\end{array}$$

where K is the quotient field of R.

But K[X] is a principal ideal domain. Therefore, $G_0(K[X])$ is isomorphic to Z and is generated by [K[X]]. K is a field. Therefore, $G_0(K)$ is isomorphic to Z and is generated by [K]. Therefore, the map on right is an iso. By assumption, the map on the left is an iso. The 5 lemma implies the middle map is onto. Done.

<u>Corollary 5.19</u>. Let k be a field. Then $G_0(k[X_1, \ldots, X_n])$ is isomorphic to Z.

<u>Corollary 5.20</u>. Let k be a field. Then $K_0(k[X_1, \ldots, X_n])$ is isomorphic to Z.

<u>Proof</u>. $k[x_1, \ldots, x_n]$ is regular. Hence, $G_0(k[x_1, \ldots, x_n])$ is isomorphic to $K_0(k[x_1, \ldots, x_n])$. Apply corollary 5.19 to obtain the result. Done.

<u>Remark</u>. Corollary 5.20 is equivalent to proving if P is a finitely generated projective $k[x_1, \ldots, x_n]$ module, then there is a finitely generated free $k[x_1, \ldots, x_n]$ module F such that $P \oplus F$ is free. It is unknown if there are finitely generated projective $k[x_1, \ldots, x_n]$ modules which are not free. For n = 0, it is trivial and for 2 it is a result of Seshadri that all finitely generated projectories are free. Bass has proved that if rank $P > n$, then P is free.

Let R be a commutative noetherian ring. Let $\widetilde{G_0(R)}$ be $G_0(R)/S$ where S is the subgroup generated by [R].

If R is also a domain, then the map of Z to G(R) given by $1 \rightsquigarrow [R]$ is split by $G(R) \longrightarrow G(K) = Z$ where $[M] \rightsquigarrow [M \otimes_R K]$ and K is the quotient field of R. Hence $G_0(R) = Z \oplus \widetilde{G_0(R)}$.

If R is not a domain, then $1 \rightsquigarrow [R]$ still gives rise to a mono which may not be split. For example, let $R = k[x]/(x^2)$ where k is a field. Then the map $Z \longrightarrow G_0(R)$ given by $1 \rightsquigarrow [R]$ does not split because $G_0(R)$ is isomorphic to Z generated by [k] and $[R] = 2[k]$ as can be seen by the sequence $0 \longrightarrow (x) \longrightarrow R \longrightarrow R/(x) \longrightarrow 0$ If R has no nilpotent elements, then $G_0(R)$ is isomorphic to $Z \oplus \widetilde{G_0(R)}$, and we obtain one splitting for each minimal prime ideal, $\underline{p}$, of R. Then $R_{R-\underline{p}}$ is a field and the map $1 \rightsquigarrow [R]$ is split by $[M] \rightsquigarrow [M \otimes_R R_{R-\underline{p}}]$. Note that in the above argument R must be noetherian. Otherwise $[R] \notin G_0(R)$.

Proposition 5.9 . Let R be a commutative noetherian ring and $s \in R$ be a non zero divisor. Then

$$\widetilde{G_0(R/(s))} \longrightarrow \widetilde{G_0(R)} \longrightarrow \widetilde{G_0(R_S)} \longrightarrow 0$$

is exact where S is the multiplicative system generated by s.

Proof. We already know by Corollary 5.15 that $G_0(R/(s)) \longrightarrow G_0(R) \longrightarrow G_0(R_S) \longrightarrow 0$ is exact. But the exact sequence of R modules $0 \longrightarrow R \xrightarrow{\hat{s}} R \longrightarrow R/(s) \longrightarrow 0$ shows that $[R/(s)] = 0$ in $G_0(R)$. Hence $\widetilde{G_0(R/(s))} \longrightarrow G_0(R) \longrightarrow G_0(R_S) \longrightarrow 0$ is exact. But $[R] \longrightarrow [R_S]$ in the map $G_0(R) \longrightarrow G_0(R_S)$. Hence $\widetilde{G_0(R/(s))} \longrightarrow \widetilde{G_0(R)} \longrightarrow \widetilde{G_0(R_S)} \longrightarrow 0$ is exact. Done.

Remark. Let k be a field of char $\neq 2$ and with $\sqrt{-1} \in k$. Define $R_n = k[x_0, \ldots, x_n]/(x_0^2 + \ldots + x_n^2 - 1)$. If k is the field of complex numbers, R_n is contained in the ring of continuous complex functions on the n-sphere, S^n. In topological K-theory it is true that $\widetilde{K^0(S^n)}$ is 0 if n is odd and is Z if n is even. Claborn and Fossum have proved that $\widetilde{G_0(R_n)}$ is 0 if n is odd and Z if n is even. The complete proof of this requires more algebraic geometry than we wish to introduce. However, we will indicate the basic approach. We note that R_n is a regular domain for $n > 0$. In R_n we have

$$x_i^2 + x_{i+1}^2 = (x_i + \sqrt{-1}\, x_{i+1})(x_i - \sqrt{-1}\, x_{i+1}).$$

Let $y_i = x_i + \sqrt{-1}\, x_{i+1}$ and $y_{i+1} = x_i - \sqrt{-1}\, x_{i+1}$. Then, since char $k \neq 2$, the y_i's generate the same ring as the x_i's. If n is odd, then there are an even number of x_i's and hence we can pair them all into y_i's getting

$$R_n = k[y_0, \ldots, y_n]/(y_0y_1 + y_2y_3 + \ldots + y_{n-1}y_n - 1).$$

If n is even, then we let $y_0 = x_0$ and pair the remaining x_i's to create y_i's as above giving

$$R_n = k[y_0, \ldots, y_n]/(y_0^2 + y_1y_2 + \ldots + y_{n-1}y_n - 1).$$

Let $R_n[y_n^{-1}]$ be R_n localized at the multiplicative system of powers of y_n. By the previous proposition we have

$$\widetilde{G_0(R_n/(y_n))} \longrightarrow \widetilde{G_0(R_n)} \longrightarrow \widetilde{G_0(R_n[y_n^{-1}])} \longrightarrow 0$$

is exact. But $R_n[y_n^{-1}]$ is isomorphic to $k[y_0, \ldots, y_{n-2}, y_n, y_n^{-1}]$. Hence, $\widetilde{G_0(R_n[y_n^{-1}]} = 0$. On the other hand $R_n/(y_n)$ is isomorphic to $R_{n-2}[y_{n-1}]$ and $\widetilde{G_0(R_{n-2}[y_{n-1}])} = \widetilde{G_0(R_{n-2})}$. Thus there is an epi $G_0(R_{n-2}) \longrightarrow G_0(R_n) \longrightarrow 0$.

If n is odd, then $R_{-1} = k$ and, therefore, $\widetilde{G_0(R_{-1})} = 0$. Thus $\widetilde{G_0}(R_n) = 0$ for all odd n.

If n is even, we note that $R_0 = k[y_0]/(y_0^2 - 1)$ is isomorphic to $k \oplus k$. Thus $G_0(R_0) = Z \oplus Z$ and $\widetilde{G_0(R_0)} = Z$. Hence, if n is even $\widetilde{G_0}(R_n)$ is cyclic.

Let X be the space in P^{n+1}, projective n + 1 space, defined by $x_0^2 + \dots + x_n^2 - x_{n+1}^2 = 0$ and let A be the subspace defined by $x_0^2 + \dots + x_n^2 = 0$. Then X - A is affine defined by $x_0^2 + \dots + x_n^2 - 1$. Then $G_0(R_n)$ is K_0 (category of coherent sheaves on X - A). In general, let G_0 (affine space) = K_0 (category of coherent sheaves on the affine space). Then we obtain the exact sequence

$$G_0(A) \longrightarrow G_0(X) \longrightarrow G_0(X - A) \longrightarrow 0.$$

Next we apply the theorem of Grothendieck.

Theorem 5.19. Let X be projective and nonsingular. Then the rank of $G_0(X)$ is the rank of the Chow ring of X.

But the ranks of the Chow rings were computed classically (cf. Hodge and Pedoe) and in particular the rank of $G_0(X - A)$ has rank $\geqslant 2$. Therefore $\widetilde{G_0(R_n)}$ is isomorphic to Z.

Chapter 6. K_0 of Graded Rings.

The purpose of this section is to prove that for R left regular $K_0(R)$ is isomorphic to $K_0(R[x])$ and to $K_0(R[x, x^{-1}])$. This theorem is false for R not regular. Bass and Murty have shown that $K_0(ZG)$ is not finitely generated for most finitely generated abelian groups G. It is known that $K_0(ZG)$ is finitely generated when G is a finite group.

Definition. A graded ring is a ring R with a decomposition of R into a direct sum of abelian groups $R = R_0 \oplus R_1 \oplus \cdots$ such that if $x \in R_n$ and $y \in R_m$ then $xy \in R_{m+n}$. We let $\overline{R}$ denote R without the grading.

An example of a graded ring is k[x] where $R_n = kx^n$ and k is any ring. Any ring R, has the trivial grading $R_0 = R$ and $R_n = 0$ $n \neq 0$.

Instead of proving that $K_0(R)$ is isomorphic to $K_0(R[x])$ if R is regular we prove the more general theorem of Serre.

Theorem 6.1. If R is a graded ring such that $\overline{R}$ is left regular then the map $R_0 \longrightarrow \overline{R}$ induces an isomorphism $K_0(R_0) \longrightarrow K_0(\overline{R})$.

Corollary 6.2. If R is left regular, then $K_0(R)$ is isomorphic to $K_0(R[x])$.

Proof. If R is regular then so is R[x].

Corollary 6.3. If R is left regular, then $K_0(R)$ is isomorphic to $K_0(R[x, x^{-1}])$.

Proof. Using the map $R[x, x^{-1}] \longrightarrow R$ given by sending x to 1 we get the sequence of maps

$$K_0(R) \longrightarrow K_0(R[x]) \longrightarrow K_0(R[x, x^{-1}]) \longrightarrow K_0(R)$$

whose composite is the identity. It remains to check that $K_0(R[x])$ to $K_0(R[x, x^{-1}])$ is onto. But R is left regular so it suffices to check that $G_0(R[x])$ to $G_0(R[x, x^{-1}])$ is onto. But this follows easily from theorem 5.13. Done with corollary 6.3 .

Proof of theorem 6.1. The maps $R_0 \longrightarrow \overline{R} \longrightarrow R/\sum_{i=1}^{\infty} R_i = R_0$ induces maps $K_0(R_0) \longrightarrow K_0(\overline{R}) \longrightarrow K_0(R_0)$ whose composite is the identity. Hence it only remains to show that the map $K_0(R_0) \longrightarrow K_0(\overline{R})$ is onto.

<u>Definition</u>. Let R be a graded ring. A <u>graded R-module</u> is an $\overline{R}$ M with a direct sum decomposition as abelian groups $M = \sum_{-\infty}^{\infty} M_n$ such that $R_n M_m \subset M_{n+m}$. Let M and N be graded R modules then a morphism f: $M \longrightarrow N$ is an $\overline{R}$ homomorphism from M to N such that $f(M_n) \subset N_n$ for all n.

Let ${}_R\underline{G}$ be the category of graded R modules M such that for each M there exists an n_0 such that $M_n = 0$ if $n < -n_0$. Let ${}_R\underline{G}^+$ be the category of graded R modules M such that $M_n = 0$ if $n < 0$.

We define two functors T: ${}_{R_0}\underline{G} \longrightarrow {}_R\underline{G}$ and Q: ${}_R\underline{G} \longrightarrow {}_{R_0}\underline{G}$.

<u>Definition</u>. Let M be a graded R module. Then the set of <u>decomposable elements of degree n of M</u>, $D_n(M)$, is $\sum_{\substack{i+j=n \\ j<n}} R_i M_j$. $D_n(M)$ is clearly a R_0 submodule of M_n.

The functor Q is defined by setting $Q_n(M)$ equal to $M_n/D_n(M)$ and $Q(M) = \coprod_n Q_n(M)$. The functor T is given by $T(M) = R \otimes_{R_0} M$ where $T(M)_n = \sum_{i+j=n} R_i \otimes_{R_0} M_j$. If $M_n = 0$ for all $n < n_0$ then $T(M)_n = 0$ for all $n < n_0$. Hence $T(M) \in {}_R\underline{G}$ when $M \in {}_{R_0}\underline{G}$.

<u>Proposition 6.4</u>. Q is right exact.

<u>Proof</u>. Let $0 \longrightarrow M' \longrightarrow M \longrightarrow M'' \longrightarrow 0$ be an exact sequence in ${}_R\underline{G}$. It is enough to show that $Q_n(M') \longrightarrow Q_n(M) \longrightarrow Q_n(M'') \longrightarrow 0$ is exact. It is clear that $D_n \longrightarrow D''_n \longrightarrow 0$ is exact. A simple chase of the following diagram gives the result:

$$\begin{array}{ccccccccc}
 & & 0 & & 0 & & 0 & & \\
 & & \downarrow & & \downarrow & & \downarrow & & \\
 & & D'_n & \rightarrow & D_n & \rightarrow & D''_n & \rightarrow & 0 \\
 & & \downarrow & & \downarrow & & \downarrow & & \\
0 & \rightarrow & M'_n & \rightarrow & M_n & \rightarrow & M''_n & \rightarrow & 0 \\
 & & \downarrow & & \downarrow & & \downarrow & & \\
 & & Q'_n & \rightarrow & Q_n & \rightarrow & Q''_n & \rightarrow & 0 \\
 & & \downarrow & & \downarrow & & \downarrow & & \\
 & & 0 & & 0 & & 0 & &
\end{array}$$

Done with proposition 6.4.

Lemma 6.5. If $M \in {}_R\underline{G}$ and $Q(M) = 0$, then $M = 0$.

Proof. Let n be the smallest n such that $M_n \neq 0$. Then

$D_n(M) = \sum_{\substack{i+j=n \\ j<n}} R_i M_j = 0$ and $Q_n(M) = M_n$. Hence all $M_n = 0$ and $M = 0$.

Done with lemma 6.5.

Theorem 6.6. If P is a projective R module in ${}_R\underline{G}$ then P is isomorphic to $TQ(P)$ and $Q(P)$ is projective. If $P \in {}_{R_0}\underline{G}$ is projective then P is isomorphic to $QT(P)$ and $T(P)$ is projective.

Proof. Let I be an index set and let $d_i \in Z$ with $d_i \geq -n_0$ all i. A free graded R module, F, on e_i is a free $\overline{R}$ module on the e_i graded by $F_n = \sum r_i e_i$ where $r_i \in R_{g_i}$ and $g_i + d_i = n$.

Let F' be a free graded R_0 module on the e_i. F'_n is a R_0 free module on those e_i with $d_i = n$.

$$T(F') = R \otimes_{R_0} F' = F.$$

$$Q_n(F) = F_n / D_n(F).$$

$$D_n(F) = \{ \textstyle\sum r_i e_i \mid r_i \in R_{g_i},\ g_i + d_i = n \text{ and } d_i < n\}.$$

Hence $Q_n(F) = F'_n$ and $TQ(F) = F$.

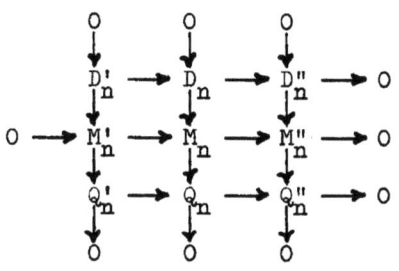

Hence, theorem 6.6. is done for free modules. The usual argument shows free modules are projective and that if P is a projective graded R module then can find P' also projective with $P \oplus P'$ free.

$Q(P \oplus P') = Q(P) \oplus Q(P')$ is free. Hence Q sends projectives to projectives.

Let $I = R_1 \oplus R_2 \oplus \dots$. I is a 2 sided R_0 ideal and R/I is isomorphic to R_0. Let P be a projective graded R module. Then

$f: P \longrightarrow P/IP = Q(P)$ is an R_0 map and Q(P) is a projective R_0 module. But $P \longrightarrow P/IP$ is onto, therefore, there is an R_0 map, g, splitting it. (Note since we have to pick a splitting we lose naturality). This extends to $h: TQ(P) = R \otimes_{R_0} Q(P) \longrightarrow P$. We claim this is an iso. Clearly $Q(R \otimes_{R_0} Q(P)) = Q(P)$ and Q(h) gives the identity map of this on Q(P). Hence Q(h) is an iso. Hence Q(coker h) = 0. By lemma 6.5. coker h = 0 and h is onto. P projective implies h splits. The same argument applied to a map splitting h shows that ker h = 0 and h is an iso. Hence P and TQ(P) are isomorphic for P projective. The other half of the theorem is clear.

Done with theorem 6.6.

Corollary 6.7. If P is a finitely generated projective in ${}_R\underline{G}$ then P is isomorphic to $R \otimes_{R_0} P'$ where P' is a finitely generated projective in $R_0\underline{G}$.

Proof. Take P' = Q(P). Done.

Since Q is right exact, Q takes finitely generated modules to finitely generated modules. T obviously has the same property.

In order to prove theorem 6.1 we will need some facts about relations between $M \in {}_R\underline{G}$ and $\overline{M}$, the $\overline{R}$ module given by forgetting the grading. The $\overline{\ }$ functor is exact.

Proposition 6.8. M is finitely generated over R if and only if $\overline{M}$ is finitely generated over $\overline{R}$.

Proof. $\Longrightarrow$ clear.

$\Longleftarrow$ Let $m_1, \ldots, m_n \in \overline{M}$ generated $\overline{M}$ over $\overline{R}$. Then each $m_i = \sum_{-\infty}^{\infty} m_{in}$ with $m_{in} \in M_n$ and almost all $m_{in} = 0$. The nonzero m_{in} generate M over R. Done with proposition 6.8.

Corollary 6.9. $\overline{R}$ left noetherian implies R is left noetherian.

Proposition 6.10. $M \in {}_R\underline{G}$ is projective if and only if $\overline{M}$ is a projective $\overline{R}$ module.

Proof. $\Longrightarrow$ M projective implies $M \oplus N = F$ with F free. Then $\overline{M} \oplus \overline{N} = \overline{F}$ and $\overline{M}$ is projective.

$\Longleftarrow$ Suppose $\overline{M}$ is projective. Choose a free module F in ${}_R\underline{G}$ and an epi $F \xrightarrow{f} M \longrightarrow 0$. Then there is a $\overline{R}$ map $\overline{g}\colon \overline{M} \longrightarrow \overline{F}$ splitting $\overline{f}$. $\overline{g}$ can fail to be a R map. Let $g_{nm}\colon M_n \longrightarrow F_m$ be the composition $M_n \longrightarrow \overline{M} \xrightarrow{g} \overline{F} \longrightarrow F_m$. Then $fg_{nn}(x) = x$ and $fg_{nm}(x) = 0$ if $n \neq m$ since $fg = \text{id}$. Define $h\colon M \longrightarrow F$ by letting $h_n\colon M_n \longrightarrow F_n$ be g_{nn}. Then $fh = \text{id}$. We claim h is an R map. Let $r \in R_k$ and $x \in M_n$. Then $g\colon rx \in M_{n+k} \rightsquigarrow (g_{n+k}, m(rx)) = (rg_{n+k}, m(x))$. Therefore $g_{n+k,n+k}(rk) = rg_{n,n}(x)$. Done with proposition 6.10.

Corollary 6.11. If $M \in {}_R\underline{G}$ then $\text{pd}_R M = \text{pd}_{\overline{R}}\overline{M}$.

Proof. Let $0 \longrightarrow B \longrightarrow F_{n-1} \longrightarrow \ldots \longrightarrow F_0 \longrightarrow M \longrightarrow 0$ be exact with F_i all free. Then

$0 \longrightarrow \overline{B} \longrightarrow \overline{F}_{n-1} \longrightarrow \ldots \longrightarrow \overline{F}_0 \longrightarrow \overline{M} \longrightarrow 0$ is exact and all $\overline{F}_i$ are free and B is projective if and only if $\overline{B}$ is. But $\text{pd}_R M \leqslant n$ if and only if B is projective and $\text{pd}_{\overline{R}}\overline{M} \leqslant n$ if and only if $\overline{B}$ is projective. Hence $\text{pd}_R M \leqslant n$ if and only if $\text{pd}_{\overline{R}}\overline{M} \leqslant n$.

Done with corollary 6.11.

Corollary 6.12. If $\overline{R}$ is left regular, $M \in {}_{R}\underline{G}$ and M is finitely generated, then M has a finite projective resolution by modules of the form $R \otimes_{R_0} P_i$ where the P_i are finitely generated projective R_0 modules.

Proof. Take a finite resolution of M by finitely generated projectives in ${}_{R}\underline{G}$. This is possible by corollary 6.11. Then each of the projectives is of the required form by corollary 6.7.

Done with corollary 6.12.

Now we have the problem of getting from ungraded modules over $\overline{R}$ to graded ones since not every $\overline{R}$ module is of the form $\overline{M}$. We form the graded ring R[t] where $R[t]_n = \sum_{i+j=n} R_i t^j$. Then $\overline{R[t]} = \overline{R}[t]$ and $R[t]_0 = R_0$. We define a functor F: ${}_{R[t]}\underline{G} \longrightarrow {}_{\overline{R}}\underline{M}$ by sending M to $\overline{M}/(t-1)\overline{M} = R[t]/(t-1) \otimes_{\overline{R[t]}} \overline{M}$.

Lemma 6.13. F is exact.

Proof. F is right exact since tensoring is right exact. Let $0 \longrightarrow M' \longrightarrow M$ be exact. Then $0 \longrightarrow \overline{M}' \longrightarrow \overline{M}$ is also exact. Hence it is enough to show that $((t-1)\overline{M}) \cap \overline{M}' = (t-1)\overline{M}'$. Let $m' = (t-1)m$. Then $m' = \sum m_i'$ and $m = \sum m_i$ with m_i' and m_i of degree i. The part of degree i of $(t-1)m$ is $tm_{i-1} - m_i$. This must equal m_i'. Thus $tm_{i-1} - m_i \in M_i'$. Let j be the smallest i such that $m_j \notin M_j'$. Then $tm_{j-1} \in M_j'$ since $m_{j-1} \in M_{j-1}'$ and, hence, $m_j \in M_j'$. This is a contradiction. Since there is an n_0 such that if $i < n_0$ $M_i = 0$ we have that $m_i \in M_i'$ for all i. Done with lemma 6.13.

Lemma 6.14. Let M be a finitely presented $\overline{R}$ module, then there is a finitely generated R[t] module N in ${}_{R[t]}\underline{G}$ such that $F(N) = \overline{M}$.

<u>Proof</u>. Let H be a free $\bar{R}$ module. If H' is any free R[t] module on the same number of generators, then F(H) = H'.

To finish the lemma it is enough to show that if f: $H_1 \longrightarrow H_0$ where H_1 and H_0 are finitely generated free $\bar{R}$ modules then there is a g: $H_1' \longrightarrow H_0'$ such that $F(H_1') = H_1$, $F(H_0') = H_0$ and F(g) = f.

Pick a basis u_i for H_1 and e_i for H_0. Then $F(u_i) = \sum a_{ij} e_j$ with $a_{ij} \in \bar{R}$.

Let $a_{ij}^{(n)}$ be the component of a_{ij} in R_n. Then there is an N such that $a_{ij}^{(n)} = 0$ for all $n \geq N$. Choose u_i of degree N to be a basis for H_1' and e_j' of degree 0 to be a basis for F_0'. Then define g by $g(u_i') = \sum t^{N-n} a_{ij}^{(n)} e_j'$. This clearly works since t goes to 1 under F. Done with lemma 6.14.

Finally we can show that the map of $K_0(R_0) \longrightarrow K_0(\bar{R})$ is onto. Let M be a finitely generated projective R module. By lemma 6.14 there is an $N \in {}_{\overline{R[t]}}\underline{G}$ such that M = F(N). Now $\overline{R[t]}$ is left regular since $\bar{R}$ is. Therefore by corollary 6.12 there is a finite projective resolution of N by modules of the form $P_i = R[t] \otimes_{R_0} P_i'$ where P_i' is a finitely generated projective R module. $F(P_i) = \bar{R} \otimes_{R_0} P_i'$ and by applying F to the resolution of N by P_i we get the exact sequence

$$0 \longrightarrow \bar{R} \otimes_{R_0} P_n' \longrightarrow \cdots \longrightarrow \bar{R} \otimes_{R_0} P_0' \longrightarrow M \longrightarrow 0 .$$

Hence $[M] = \sum (-1)^i [F(P_i)] = \sum (-1)^i [\bar{R} \otimes_{R_0} P_i']$ in $K_0(\bar{R})$ and this clearly comes from $K_0(R_0)$. Done with theorem 6.1.

Chapter 7. Spec(R) and H(R)

Definition. Let R be a commutative ring. We define **spec(R)** to be the set of prime ideals of R and **m-spec(R)** to be the set of maximal ideals of R. If $f: R \to R'$ is a homomorphism and p a prime ideal of R' then the map $f^*: \mathrm{spec}(R') \to \mathrm{spec}(R)$ given by $f^*(p) = f^{-1}(p)$ makes spec into a contravariant functor. m-spec is not a functor in this sense, for if $f: Z \to Q$ is the standard imbedding of the integers into the rationals then 0 is a maximal ideal of Q but $f^{-1}(0) = 0$ is not a maximal ideal of Z.

If R is noncommutative, a modified definition is still possible using a formulation due to Procesi. We call a proper 2 sided ideal p of R prime if for every pair of 2 sided ideals a and b $ab \subset p$ implies $a \subset p$ or $b \subset p$. This agrees with the usual definition if R is commutative. Again we set spec R = set of primes and m-spec R = set of maximal 2 sided ideals of R. (Note that maximal 2 sided ideals will be prime cf. lemma 7.2.)

In the noncommutative case spec(R) is not a functor, as the following example shows. Let k be any field and $M_2(k)$ be the 2 by 2 matrices over k. $M_2(k)$ is simple and hence 0 is the only prime ideal. Let B be the ring of matrices of the form $\binom{**}{0*}$ and f the injection of B into $M_2(k)$. Then $0 = f^{-1}(0)$ but 0 is not a prime ideal of B. For if a is all matrices of the form $\binom{0+}{00}$, a is a 2-sided ideal of B and $a^2 = 0$ and $a \neq 0$. Hence $f^{-1}(0)$ is not prime in B.

If a is a 2 sided ideal of R, set $F(a) = \{p \in \mathrm{spec}(R) \mid p \supset a\}$ and $F_m(a) = \{m \in \text{m-spec}(R) \mid m \supset a\}$. We make spec(R) (respectively m-spec(R)) into a topological space with $F(a)$ $(F_m(a))$ as closed sets.

Note first that $F(R) = \emptyset$ and $F(0) = \text{spec } R$. If $\underline{a}_i$ is a collection of 2 sided ideals of R, then $F(\sum \underline{a}_i) = \bigcap_i F(\underline{a}_i)$. To account for finite unions we prove:

$$F(\underline{a}) \cup F(\underline{b}) = F(\underline{ab}) = F(\underline{a} \cap \underline{b}) .$$

If $\underline{x} \subset \underline{y}$ are 2 sided ideals, then $F(\underline{x}) \supset F(\underline{y})$. It follows that $F(\underline{ab}) \supset F(\underline{a} \cap \underline{b}) \supset F(\underline{a}) \cup F(\underline{b})$. But if $\underline{p} \supset \underline{ab}$ and $\underline{p}$ is prime, then $\underline{p} \supset \underline{a}$ or $\underline{p} \supset \underline{b}$. Therefore $\underline{p} \in F(\underline{a})$ or $\underline{p} \in F(\underline{b})$. Thus $F(\underline{ab}) \subset F(\underline{a}) \cup F(\underline{b})$. Hence $F(\underline{a}) \cup F(\underline{b}) = F(\underline{ab})$. Spec(R) with this topology, called the Zariski topology, is a topological space. The open sets are $W(\underline{a}) = \{\underline{p} \in \text{spec}(R) \mid p \not\supset \underline{a}\}$ for some 2 sided ideal $\underline{a}$ of R.

One easily shows that if $f: R \longrightarrow R'$ is a homomorphism, then $f^*: \text{spec}(R') \longrightarrow \text{spec}(R)$ is continuous.

The topology on m-spec R coincides with the subspace topology from spec R. In general m-spec R is neither open nor closed in spec R.

The identities on F listed above translate immediately to the following statements about W:

1) $W(0) = \emptyset \quad W(R) = \text{spec}(R)$,

2) $W(\sum \underline{a}_i) = \bigcup W(\underline{a}_i)$,

3) $W(\underline{ab}) = \quad W(\underline{a}) \cap W(\underline{b}) = W(\underline{a} \cap \underline{b})$, and

4) $\underline{a} \subset \underline{b}$ implies $W(\underline{a}) \subset W(\underline{b})$.

<u>Lemma 7.1</u>. $W(\underline{a}) = \text{spec}(R)$ if and only if $\underline{a} = R$.

<u>Proof</u>. $\underline{a} \neq R$ implies $\underline{a} \subset \underline{m}$ where $\underline{m}$ is a maximal 2 sided ideal (recall that all rings have 1). If we could prove $\underline{m} \in \text{spec}(R)$ we would be done. We generalize this to the following:

Lemma 7.2. Let S be a (not necessarily central) multiplicatively closed subset of R not containing 0. If $\underline{a}$ is a 2 sided ideal maximal with respect to $\underline{a} \cap S = \emptyset$, then $\underline{a}$ is prime.

Proof. Let $\underline{bc} \subset \underline{a}$, $\underline{b} \not\subset \underline{a}$ and $\underline{c} \not\subset \underline{a}$. Then $(\underline{a} + \underline{b}) \cap S \neq \emptyset$ and $(\underline{a} + \underline{c}) \cap S \neq \emptyset$. Say $s \in (\underline{a} + b) \cap S$ and $t \in (\underline{a} + \underline{c}) \cap S$. Then $st \in S$ and $st \in \underline{a} + \underline{bc} = \underline{a}$. This is a contradiction. Therefore, $\underline{b} \subset \underline{a}$ or $\underline{c} \subset \underline{a}$ and $\underline{a}$ is prime. Done with lemma 7.2.

Setting $S = \{1\}$ in lemma 7.1 we see that $\underline{m}$ is prime.

Done with lemma 7.1.

Corollary 7.3. spec(R) and m-spec(R) are quasi compact (that is compact but not necessarily Hausdorff).

Proof. Let $\text{spec}(R) = \bigcup_i U_i$ where U_i are open. Set $U_i = W(\underline{a}_i)$. Then $\text{spec}(R) = \bigcup_i W(\underline{a}_i) = W(\sum \underline{a}_i)$. Hence $\sum \underline{a}_i = R$. Therefore, $1 = \sum a_i$ with $a_i \in \underline{a}_i$ almost all 0. Say $a_1, \ldots, a_n$ are all the nonzero ones. Then $\underline{a}_1 + \ldots + \underline{a}_n = R$ (since 1 is in it) and $\text{spec}(R) = \bigcup_1^n W(\underline{a}_i)$. The same proof works for m-spec. Done.

Suppose $W(\underline{a}) = \emptyset$ for $\underline{a}$ a 2 sided ideal $\underline{a}$. Then $F(\underline{a}) = \text{spec}(R)$. That is every prime ideal of R contains $\underline{a}$. Hence

$W(\underline{a}) = \emptyset$ if and only if $\underline{a} \subset \bigcap \underline{p}$

where $\underline{p}$ runs over all primes of R.

Proposition 7.4. $\bigcap \underline{p}$ where $\underline{p}$ runs over all primes of R is a nil ideal.

Proof. Suppose not. Then pick $s \in \bigcap \underline{p}$ which is not nilpotent. Then $S = \{1, s, s^2, \ldots\}$ is a multiplicative system not containing 0. Hence there exist a prime ideal $\underline{p}$ not containing s. Hence $s \notin \bigcap \underline{p}$. contradiction.

Corollary 7.5. If R is commutative, $\bigcap \underline{p}$ where $\underline{p}$ runs over all primes of R is the set of all nilpotent elements.

Corollary 7.6. If R is commutative:

1) $W(\underline{a}) = \emptyset$ in spec (R) if and only if $\underline{a}$ is nil.
2) $W(\underline{a}) = \emptyset$ in m-spec (R) if and only if $\underline{a} \subset J(R)$, the Jacobson radical of R.

From now on we assume R is commutative. If $\underline{p}$ is a prime ideal of R, $R_{\underline{p}}$ denotes R_S where $S = R - \underline{p}$ is a multiplicative system. If P is a finitely generated projective R module, then the $R_{\underline{p}}$ module $P\underline{p} = P \otimes_R R\underline{p}$ is a free $R\underline{p}$ since $R\underline{p}$ is a local ring.

Definition. $\underline{r_P(\underline{p})}$ = rank of $P\underline{p}$ over $R\underline{p}$ = the minimal number of generators of $R\underline{p}$ as a $R\underline{p}$ module = dimension over $R\underline{p}/\underline{p}_p$ of $P\underline{p}/\underline{p}_p P\underline{p}$.

r_p: spec (R) $\longrightarrow$ Z. The immediate objective is to prove that r_p is continuous where Z is given the discrete topology.

Definition. If M is an R module, the support of M, supp M, is $\{\underline{p} \in \text{spec}(R) | M_{\underline{p}} \neq 0\}$.

Lemma 7.7. If M is a finitely generated R module, then supp M is closed in spec(R). In fact supp M = F(annihilator of M).

Proof. Let $m_1, \ldots, m_n$ generate M. $M_{\underline{p}} = 0$ implies that for all $m \in M$ there exists $s \notin \underline{p}$ such that $sm = 0$. Hence there exists $s_i \notin \underline{p}$ such that $s_i m_i = 0$. Let $s = \prod s_i$. Then $sM = 0$. Hence $M_{\underline{p}} = 0$ if and only if there exists $s \notin \underline{p}$ with $sM = 0$. But $sM = 0$ if and only if $s \in$ annih(M) so $M_p = 0$ if $\underline{p} \not\supset$ annih(M). Therefore $M_{\underline{p}} \neq 0$ if and only if $\underline{p} \supset$ annih(M). Hence, supp M is closed. Done.

Remarks. 1) The lemma is clearly false if M is not finitely generated. Let R = Z and I a set of integers. $M = \coprod_{n \in I} Z/nZ$. Then

Annih(M) = $\{m|n|m$ for all $m \in I\} = \{0\}$ if I is infinite.

$M_{\underline{p}} = 0$ if and only if $p \nmid n$ for all $n \in I$.

2) If A is an R algebra and M is a finitely generated A module. Then the lemma is still true where we use the R annihilator of M, the proof being exactly the same.

Let A be an R algebra finitely generated as an R module (or "finite R algebra" for short) such that $A_{\underline{p}} \neq 0$ for all $\underline{p} \in \mathrm{spec}(R)$, and P a finitely generated projective A module such that $P_{\underline{p}}$ is free for all $\underline{p} \in \mathrm{spec}(R)$. Let $r_P(\underline{p})$ = the number of free generators of $P_{\underline{p}}$.

<u>Theorem 7.8</u>. r_P: spec(R) $\longrightarrow$ Z is continuous where Z is given the discrete topology.

We begin by showing that $r_P(\underline{p})$ is really the minimal number of generators of $P_{\underline{p}}$.

<u>Lemma</u>. Let R be a commutative ring, $A \neq 0$ an R algebra finitely generated as an R-module, and let P be a free A-module on n generators. Then P cannot be generated by fewer than n elements.

<u>Proof</u>. If P could be generated by $k < n$ elements there would be an epimorphism $A^k \longrightarrow A^n = P$. Let m be a maximal ideal of R with $A/mA \neq 0$. Then $(A/mA)^k \longrightarrow (A/mA)^n$ but there are finitely generated vector spaces over the field R/m so $k \geq n$. To see that such a m exists, note that mA = A implies $Am_mA_m = A_m$ and so, by Nakayama's lemma $A_m = 0$. If this is so for all m, then A = 0.

<u>Proof of Theorem 7.8</u>. Suppose $r_P(\underline{p}) = n$. Let F be a free A module on n generators. Then there exists an iso f: $F_{\underline{p}} \longrightarrow P_{\underline{p}}$. Since P is finitely presented, there is an A map g: F $\longrightarrow$ P such that

$$\begin{array}{ccc} F_{\underline{p}} & \xrightarrow{g} & P_{\underline{p}} \\ & {\scriptstyle \hat{s}} \searrow \quad \nearrow {\scriptstyle f} & \\ & F_{\underline{p}} & \end{array}$$

commutes.

Let M = coker g. Since $F_{\underline{p}} \longrightarrow P_{\underline{p}} \longrightarrow M_{\underline{p}} \longrightarrow 0$ is exact we conclude that $M_{\underline{p}} = 0$. That is, $\underline{p} \notin$ supp M, a closed subset of spec R. Hence, there exists an open set U containing $\underline{p}$ such that $M_{\underline{q}} = 0$ for all $\underline{q} \in U$.

When $\underline{q} \in U$, $F_{\underline{q}} \longrightarrow P_{\underline{q}} \longrightarrow M_{\underline{q}} \longrightarrow 0$ is exact and $M_{\underline{q}} = 0$ implies P can be generated by n elements. Therefore $r_P(\underline{q}) \leq n$ for all $\underline{q} \in U$.

Now let $P \oplus Q = F$ where F is a free A module on N generators. Then $P_{\underline{p}} \oplus Q_{\underline{p}} = F_{\underline{p}}$, so $Q_{\underline{p}}$ is free on N - n generators or $r_Q(\underline{p}) = N - n$. As before there is an open set V containing $\underline{p}$ such that if $\underline{q} \in V$ then $r_Q(\underline{q}) \leq N - n$. $\underline{p} \in U \cap V$ so that the open set $U \cap V \neq \emptyset$. It is clear that if $\underline{p} \in U \cap V$, then $r_P(\underline{p}) = n$. Therefore r_P is continuous. Done.

The next theorem gives a complete answer to the question of when is it true that $A_{\underline{p}} \neq 0$ for all $\underline{p} \in$ spec (R) under the assumption that A is a finitely generated R module.

<u>Theorem 7.9</u>. Let A be an R-algebra finitely generated as a module and let $\underline{a}$ be the kernel of $R \longrightarrow A$. Then $A_{\underline{p}} \neq 0$ for all $\underline{p} \in$ spec (R) if and only if $\underline{a}$ is a nil ideal.

<u>Proof</u>. $\Longrightarrow$ If $\underline{p} \in$ spec (R), then $\underline{a} \not\subset \underline{p}$ implies there exists $s \notin \underline{p}$ and $s \in \underline{a}$. Hence sA = 0. Hence $A_{\underline{p}} = 0$. Therefore $\underline{a} \subset \bigcap \underline{p}$ and $\underline{a}$ is nil by proposition 7.4.

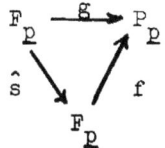

$\Longleftarrow$ If $A_{\underline{p}} = 0$ and A is generated by $a_1, \ldots, a_n$ then there exists $s_i \in R - \underline{p}$ such that $s_i a_i = 0$. Let $s = \prod s_i$. Then sA = 0

implies $s \in \underline{a}$ and $s \notin \underline{p}$ which is prime. Since $\underline{a}$ is nil, $\underline{a} \subset \cap\, \underline{p}$. Therefore $s \in \underline{p}$. This is a contradiction. Hence, $A_{\underline{p}} \neq 0$ for all $\underline{p} \in$ spec R. Done.

<u>Definition</u>. Let R be a commutative ring. Then <u>H(R)</u> is the group of continuous functions spec (R) $\longrightarrow$ Z where Z is given the discrete topology.

Note that H(R) is a functor, for if f: R $\longrightarrow$ R' is a ring homomorphism and $g \in$ H(R) then gH(f): spec R' $\xrightarrow{f^*}$ spec R $\xrightarrow{g}$ Z is a continuous function from spec R' to Z. The function r gives a homomorphism $K_0(R) \longrightarrow H(R)$ via $r([P]) = r_P$. To show that r indeed defines a homorphism we must show $r_{P \oplus Q} = r_P + r_Q$. But this is obvious since $(P \oplus Q)_{\underline{p}} = P_{\underline{p}} \oplus Q_{\underline{p}}$ for all $\underline{p} \in$ spec (R).

<u>Corollary 7.10</u>. The homomorphism r: $K_0(R) \longrightarrow H(R)$ is a natural transformation of functors.

<u>Proof</u>. Suppose f: R $\longrightarrow$ R'. We must show the commutativity of the following diagram:

$$\begin{array}{ccc} K_0(R) & \xrightarrow{\ r\ } & H(R) \\ {\scriptstyle K_0(f)}\downarrow & & \downarrow{\scriptstyle H(f)} \\ K_0(R') & \xrightarrow{\ r\ } & H(R') \end{array} .$$

We start with $[P] \in K_0(R)$. If $\underline{p} \in$ spec R', then chasing the diagram in both directions leads to the proposed equality:

$$r_{R' \otimes_R P}(\underline{p}) = r_P(f^{-1}(\underline{p})).$$

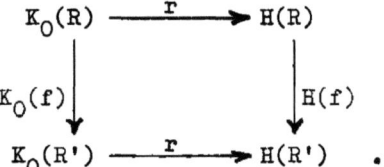

That is if $q = f^{-1}(\underline{p})$, we must show that $P_{\underline{q}}$ and $(R' \otimes_R \bar{P})_{\underline{p}}$ are free on the same number of generators. This is accomplished by the following lemma:

Lemma 7.11. $(R' \otimes_R P)_{\mathfrak{p}}$ is isomorphic to $R'_{\mathfrak{p}} \otimes_{R_{\mathfrak{q}}} P_{\mathfrak{q}}$.

Proof. If M is an $R_{\mathfrak{p}}$ module, then $\mathrm{Hom}_{R'_{\mathfrak{p}}}((R' \otimes_R P)_{\mathfrak{p}}, M)$ is isomorphic to $\mathrm{Hom}_{R'}(R' \otimes_R P, M)$ is isomorphic to $\mathrm{Hom}_R(P, M)$ is isomorphic to $\mathrm{Hom}_{R_{\mathfrak{q}}}(P_{\mathfrak{q}}, M)$ is isomorphic to $\mathrm{Hom}_{R'_{\mathfrak{p}}}(R'_{\mathfrak{p}} \otimes_{R_{\mathfrak{q}}} P_{\mathfrak{q}}, M)$.

Hence both sides of lemma 7.11 represent the same functor. Hence they are isomorphic. Done.

If A is a finite projective (as a module) R algebra and A' a finite projective R' algebra and g an algebra map $g: A \longrightarrow A'$ over f (that is

$$\begin{array}{ccc} A & \xrightarrow{g} & A' \\ \uparrow & & \uparrow \\ R & \xrightarrow{f} & R' \end{array}$$

commutes).

then r defines a natural transformation of functors $K_0(A) \longrightarrow H(R)$. That is

$$\begin{array}{ccc} K_0(A) & \xrightarrow{r} & H(R) \\ {\scriptstyle K_0(g)}\downarrow & & \downarrow{\scriptstyle H(f)} \\ K_0(A') & \longrightarrow & H(R') \end{array}$$

commutes.

The proof is exactly the same as in the theorem.

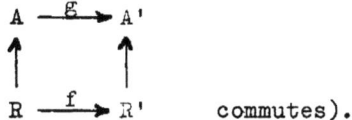

If $f \in H(R)$, then for any $n \in Z$, $U_n = f^{-1}(n)$ is a subset of spec (R) which is both open and closed. Furthermore, since $\bigcup_n U_n$ = spec (R) and spec (R) is quasi compact, a finite number of the U_i cover spec (R). Of course, $U_n \cap U_m = \emptyset$ if $m \neq n$ and only a finite number of U_i are nonempty. Hence, we have proved that

1) any element $f \in H(R)$ has finite image and

2) any $f \in H(R)$ induces a decomposition of spec R into a disjoint union of open and closed sets.

Now we analyze what the decompositions in 2) say about the ring R.

Theorem 7.12. Let R be a commutative ring. There is a one-one correspondence between:

1) direct factors of R,

2) idempotents of R, and

3) closed and open subsets of spec (R).

Proof. The correspondence between 1) and 2) is quite standard. If $R = R' \oplus R''$, then the idempotent $(1, 0) = e'$ determines this decomposition. For $R' = Re'$ and $R'' = R(1 - e')$ and e' is uniquely determined since it is the unit of R'. Conversely if $e \in R$ is an idempotent, then

*) $R = Re + R(1 - e)$.

Since $(1 - e)$ is idempotent and $e(1 - e) = 0$, we conclude that $Re \cap R(1 - e) = 0$. Hence, the sum in *) is direct. As before e is the unit of Re. This finishes the correspondence between 1) and 2).

If e is any idempotent of R, $R = Re + R(1 - e)$. Hence, spec $(R) = W(R) = W(Re) \cup W(R(1 - e))$ and $W(Re) \cap W(R(1 - e)) =$ $= W(Re\, R(1 - e)) = W(0) = 0$.

$W(Re) \cap W(Re') = W(ReRe') = W(Ree') = \emptyset$

if and only if Ree' is nil. Hence ee' is nilpotent and idempotent. Hence $ee' = 0$. Conversely if $ee' = 0$, then $W(Re) \cap W(Re') = \emptyset$. Hence if e and e' are idempotent then $W(Re) \cap W(Re') = \emptyset$ if and only if $ee' = 0$.

We claim that if e and e' are idempotents then W(Re) = W(Re') if and only if e = e'. If W(Re) = W(Re') then $W(Re) \cap W(R(1 - e')) = \emptyset$. Hence, e(1 - e') = 0. Hence e = ee'. By symmetry e' = ee'. Therefore, e = e'. Thus, the sets W(Re) are distinct for distinct idempotents.

Since the open set W(Re) has an open set, W(R(1 - e)), as compliment, W(Re) is also closed. Therefore every idempotent gives rise to distinct closed and open subsets of spec (R).

Let spec (R) = $U \cup V$ where U and V are disjoint open subsets. Write U = $W(\underline{a})$ and V = $W(\underline{b})$. $W(\underline{a}) \cup W(\underline{b})$ = spec (R) implies $\underline{a} + \underline{b} = R$. $W(\underline{a} \cap \underline{b}) = W(\underline{a}) \cap W(\underline{b}) = \emptyset$ implies $\underline{a} \cap \underline{b} = \underline{c}$ is a nil ideal of R. But $R/\underline{c} = \underline{a}/\underline{c} \oplus \underline{b}/\underline{c}$. Write $\underline{a}/\underline{c} = R/\underline{c}^{\overline{e}}$ where $\overline{e}$ is an idempotent of $R/\underline{c}$. If we could lift $\overline{e}$ to an idempotent e of R we would be done. Then $\underline{a} = Re + \underline{c}$, U = $W(\underline{a})$ = W(Re), and V = $W(\underline{b})$ = W(R(1 - e)). Hence, the correspondence between 2) and 3) would be done. We finish theorem 7.12 by

<u>Lemma 7.13</u>. If R is a ring and $\underline{c}$ is a nil ideal of R, then any idempotent $\overline{e}$ of $R/\underline{c}$ lifts to an idempotent e of R.

<u>Proof</u>. In the natural projection $R \twoheadrightarrow R/\underline{c}$ suppose that $x \rightsquigarrow \overline{e}$. Let Z[t] be the polynomial ring in one variable over Z and consider

$$\begin{array}{ccc} Z[t] & \xrightarrow{f} & R \\ t & \rightsquigarrow & x \end{array}$$

Suppose A = image of f and $J = A \cap \underline{c}$. Then A is a noetherian ring and J is a nil ideal of A. Therefore, J is nilpotent. Since $x^2 \equiv x(J)$, there is a $y \in A$ such that $y \equiv x(J)$ and $y^2 = y$. y is the lifted idempotent. Done with lemma 7.13 and theorem 7.12.

Corollary 7.14. If $R = R' \oplus R''$, then spec(R) is a disjoint union of spec(R') and spec(R"), where spec(R') and spec(R") are identified with open and closed subsets of spec(R) in a natural way. The functor spec( ): Rings $\longrightarrow$ Topological spaces converts finite products to disjoint sums.

Remark. If R has no nil ideals, then the proof of theorem 7.12 goes through even in the noncommutative case when we consider only central idempotents. Otherwise the theorem is false as the following example shows: Let k be a field and $A \subset M_2(k)$ be all matrices of the form $\begin{pmatrix} * & * \\ 0 & * \end{pmatrix}$, A is a ring with unit and $\underline{a} = \{\begin{pmatrix} 0 & * \\ 0 & 0 \end{pmatrix}\}$ is a 2 sided ideal with $\underline{a}^2 = 0$. Hence $\underline{a}$ is contained in every prime ideal. $A/\underline{a} \cong k \oplus k$ but nontrivial idempotents of $k \oplus k$ do not lift back to central idempotents of A. The center of $A = \{\begin{pmatrix} a & 0 \\ 0 & a \end{pmatrix} | a \in k\}$. Hence the only central idempotents of A are 0 and 1. $\underline{p}_1 = \{\begin{pmatrix} 0 & * \\ 0 & * \end{pmatrix}\}$ and $\underline{p}_2 = \{\begin{pmatrix} * & * \\ 0 & 0 \end{pmatrix}\}$ are the only prime ideals of A. But they are maximal. Hence spec(A) has the discrete topology and A does not give rise to a non trivial direct sum decomposition.

Now we want to give the connection between spec(R) and spec($R/\underline{a}$).

Theorem 7.15. Let R be a commutative ring and $\underline{a}$ be an ideal of R. Then the natural map $f: R \longrightarrow R/\underline{a}$ induces a homeomorphism $f^*: \text{spec}(R/\underline{a}) \longrightarrow \text{spec}(R)$ of spec($R/\underline{a}$) onto $F(\underline{a})$.

Proof. Inverse images of all prime ideals of $R/\underline{a}$ are exactly the prime ideals of R which contain $\underline{a}$. These form $F(\underline{a})$. Hence the map f^* is one-one and onto $F(\underline{a})$. It remains to show that the topology induced on $F(\underline{a})$ is the subspace topology from spec(R).

Let A be a closed subset of $F(\underline{a})$. Then $A = F(\underline{a}) \cap F(I)$ where I is an ideal of R. Hence, $A = F(\underline{a} + I)$ and $\underline{a} + I \supset \underline{a}$. Therefore,

$A = f^*(\underline{p} \in \mathrm{spec}(R/\underline{a})/\underline{p} \supset \underline{a} + I/\underline{a}\}$, and so A is closed in the topology determined by f^*. Conversely, if A is closed in the f^* topology, then $A = f^*(F(\underline{c}))$ where $\underline{c}$ is an ideal of $R/\underline{a}$. If $\underline{b} = f^{-1}(\underline{c})$, then $A = \{\underline{p} \in \mathrm{spec}(R) | \underline{p} \supset \underline{b}\} = F(\underline{b})$. Then $A = F(\underline{a}) \cap F(\underline{b})$ and A is a closed subset of $F(\underline{a})$ in the subspace topology. Done.

From now on we will identify spec(R/I) with F(I) for any ideal I, of a commutative ring R.

Next we want to investigate the rank map $r: K_0(R) \longrightarrow H(R)$. We will prove that r is onto. Suppose $f \in H(R)$. Then $U_n = f^{-1}(n)$ give a set of disjoint open and closed subsets of spec(R) almost all of which are empty and with $\bigcup_n U_n = \mathrm{spec}(R)$. By theorem 7.12 we may write $U_i = W(Re_i)$ for e_i a uniquely determined idempotent of R. Since $U_i \cap U_j = \emptyset$ for $i \neq j$ we conclude that $W(Re_ie_j) = \emptyset$ and that $e_ie_j = 0$ for $i \neq j$. $\bigcup_n U_n = \mathrm{spec}(R)$ implies $\sum Re_i = R$.

But $\sum Re_i = R(\sum e_i)$. For if e and e' are othogonal idempotents of R, $R(e + e') = Re + Re'$ since $r = ae + be'$ implies $r = r(e + e')$. By induction, we conclude $\sum Re_i = R(\sum e_i)$. Since $\sum e_i$ is idempotent, $\sum e_i = 1$. Let $Re_i = R_i$.

Assume first that f is non negative. Let F_i be a free R_i module on i generators. (Recall that $f|_{U_i} \equiv i$). If $P = F_1 \oplus \ldots \oplus F_n$ where $U_j = \emptyset$ if $j > n$, then P is a finitely generated projective R module. By corollary 7.14 $\mathrm{spec}(R) = \coprod \mathrm{spec}\, R_i$ where $\mathrm{spec}\, R_i = U_i$. If $\underline{p} \in U_i$ let $\underline{\underline{p}}$ be the corresponding prime ideal of R_i. That is, $\underline{\underline{p}} = f_i^{-1}(\underline{p})$ where f_i is the natural projection $f_i: R \longrightarrow R_i$.

<u>Claim</u>. 1) $R_{\underline{p}} = (R_i)_{\underline{p}}$

For $\underline{p} = R_1 \oplus \cdots \oplus \underline{p} \oplus R_{i+1} \oplus \cdots \oplus R_n$. Hence

$R-\underline{p} = 0 \oplus \cdots \oplus (R_i-\underline{p}) \oplus 0 \oplus \cdots \oplus 0$. The

claim follows easily from this.

<u>Claim</u>. 2) $P_{\underline{p}} = (F_i)_{\underline{p}}$ is free on i generators.

We have the composite $R \longrightarrow R_i \longrightarrow (R_i)_{\underline{p}} = R_{\underline{p}}$. The first change of rings identifies P with F_i; the next change of rings identifies P with $(F_i)_{\underline{p}}$. Hence the composite identifies P with $P_{\underline{p}}$ by claim 1).

Claim 2) shows that if $\underline{p} \in U_i$, $r_P(\underline{p}) = i = f(\underline{p})$. Furthermore, the module, P, does not depend on the choice of the disjoint open closed sets in the cover, $V_1, \ldots, U_n$. For, if $W_1, \ldots, W_n$ is a system of open and closed sets satisfying: $W_i \quad W_j = \emptyset$ if $i \neq j$ and $\cup W_i = \operatorname{spec}(R)$, then $\operatorname{spec}(R) = \bigcup_{i,j} U_i \cap W_j$. If $R = \coprod R_{i,j}$ is the corresponding direct sum decomposition of R, then

a) $R = \sum R_{ij}$ and

b) $F_i = \sum_j F_{ij}$ where F_{ij} is free over R_{ij} on i generators.

Hence $U_i \cap W_j$ give rise to the module P. By symmetry, the W_j give rise to the module P.

Hence the correspondence $f \rightsquigarrow [P]$ associate to every positive valued $f \in H(R)$ an element of $K_0(R)$. We denote [P] by n(f).

<u>Lemma 7.16</u>. $n(f + g) = n(g)$ where f and g are non negative functions.

<u>Proof</u>. $\operatorname{spec}(R) = U_1 \cup \cdots \cup U_n$ for f and

$\operatorname{spec}(R) = W_1 \cup \cdots \cup W_n$ for g.

Then the open and closed sets $U_i \cap W_j$ form a decomposition of spec(R) for f + g. If P_f is the module associated to f and P_g is the module associated to g, then $P_f \oplus P_g = P_{f+g}$, the module associated to f + g, since both sides are free on $f(\underline{p}) + g(\underline{p})$ generators when localized at $\underline{p} \in U_i \cap W_j$. Hence, $[P_f] + [P_g] = n(f) + n(g) = [P_f \oplus P_g] = [P_{f+g}] = n(f + g)$. Done.

Now if $f \in H(R)$ is arbitrary we may write $f = f^+ - f^{-1}$ where f^+ and f^- are non negative functions in H(R). We define $n(f) = n(f^+) - n(f^-)$. If $f = g^+ - g^-$ is another such decomposition of f, then $f^+ + g^- = g^+ + f^-$. Hence $n(f^+) - n(f^-) = n(g^+) - n(g^-)$. That is, n(f) is well-defined. It is obvious by lemma 7.16 that n is still an additive function from H(R) to $K_0(R)$.

Theorem 7.17. Let R be a commutative ring. The map $r\colon K_0(R) \to H(R)$ is a split epi. It is split by $n\colon H(R) \to K_0(R)$ and n is a natural transformation of functors.

Proof. It is clear that rn(f) = f. To prove that n is natural we must show that if $h\colon R \to R'$ is a homomorphism of commutative rings then the diagram

$$\begin{array}{ccc} H(R) & \xrightarrow{H(h)} & H(R') \\ {\scriptstyle n}\downarrow & & \downarrow \\ K_0(R) & \xrightarrow{K_0(h)} & K_0(R') \end{array}$$ commutes.

That is we must show that if $f \in H(R)$ then $[P_f \otimes_R R'] = [P_{h^*(f)} \otimes_R R']$. This is clear since tensoring is additive and sends free modules on n generators over R to free modules on n generators over R'. Done.

Definition. $\tilde{K}_0(R) = \ker r$.

Corollary 7.18. Let R be a commutative ring, then $K_0(R)$ and $H(R) \oplus \tilde{K}_0(R)$ are naturally isomorphic.

Chapter 8. Picard Group and the Determinant

Let R be a commutative ring. Recall from chapter 1 that Pic(R) is the set of isomorphism classes of finitely generated projective modules of rank 1 over R. Pic(R) forms a group under tensor product.

The map i: Pic(R) $\longrightarrow K_0(R)$ given by $[P] \rightsquigarrow [P]$ is not a homomorphism since rank $(P \otimes_R Q)$ is 1 but rank $P \oplus Q$ is 2 when P and Q have rank 1. We could change this to i': Pic(R) $\longrightarrow K_0(R)$ by $[P] \rightsquigarrow [P] - [R]$. Then the right hand side always has rank 0. But i' is not always a homomorphism. If R is a commutative noetherian ring with Krull dimension ≤ 1 then i' is a homomorphism. If R is a Dedekind ring then i' is an iso of Pic(R) with $\widetilde{K}_0(R)$.

The goal of this chapter is to construct a natural homomorphism det: $K_0(R) \longrightarrow$ Pic(R) such that $(\det)\cdot i = 1_{\mathrm{Pic}(R)}$ and $(\det)\cdot i' = 1_{\mathrm{Pic}(R)}$ for R a commutative ring. This will show immediately that i and i' are one-one. That is if P and Q are projective rank 1 and [P] = [Q] in $K_0(R)$ then P is isomorphic to Q. For the noncommutative case this is false. That is, there is a ring R and a projective module P such that $P \oplus R$ is isomorphic to $R \oplus R$ but P is not isomorphic to R. R can be taken to be Z[G], the integral group ring of the generalized quaternion group of order 32. Bass has shown that for R = Z[G] where G is any finite group if $P \oplus F$ is isomorphic to $Q \oplus F$ where P and Q are finitely generated projective R modules and F a finitely generated free R module, then $P \oplus R$ is isomorphic to $Q \oplus R$.

<u>Definition</u>. Let R be a commutative ring. Then <u>an anti-commutative graded R algebra</u>, A, is a graded ring which is an R algebra and such

that $x \in A_m$ and $y \in A_n$ implies $xy = (-1)^{mn}yx$. A is <u>strictly anti-commutative</u> if $x \in A_{2n+1}$ implies $x^2 = 0$. (Note: anti commutative implies strictly anti commutative if 2 is not a zero divisor in A.)

We have the functor S from the category of strictly anti-commutative graded R algebras to the category of R modules given by $A \rightsquigarrow A_1$. Let Λ be the left adjoint of S. (Note: Λ exists by the adjoint functor theorem. We give a more explicit construction below.)

<u>Definition</u>. Let M be an R module. Then $\Lambda(M)$ is the <u>exterior algebra of M</u>.

If F is a free R module on $\{e_i\}$ $i \in I$ we construct $\Lambda(F)$ as follows. First linearly order I. Then $\Lambda(F)_n$ is a free R module on $\{e_{i_1, \ldots, i_n}\}$ where $i_1 < i_2 < \ldots < i_n$. $\Lambda(F)_0$ is free on e. We define the product on the basis elements by

$$e_{i_1, \ldots, i_n} e_{j_1, \ldots, j_m} \begin{cases} 0 \text{ if } i_e = j_k \text{ for any l and k} \\ \sum \epsilon e_{g_1, \ldots, g_{m+n}} \text{ where the } g_l\text{'s are} \end{cases}$$

the i_k j_m

in order and $\epsilon = \operatorname{sgn} \begin{pmatrix} i_1, \ldots, i_n, j_1, \ldots, j_n \\ g_1, \ldots, g_n, \ldots\ldots, g_{n+m} \end{pmatrix}$.

$ee_{i_1} \cdots\cdots e_{i_n} = e_{i_1} \cdots\cdots e_{i_n}$.

It is clear that this multiplication is associative and strictly anticommutative. $F \longrightarrow \Lambda(F)_1$ is universal for maps of F into the grade one part of a strictly anti commutative R algebra. If $f\colon F \to G_1$, then $\overline{f}(e_{\alpha_1, \ldots, \alpha_n}) = f(e_{\alpha_1}) \ldots . f(e_{\alpha_n})$ is the desired map.

For any R module G let $i\colon G \to \Lambda(G)$, be the universal map of G into the grade one part of a strictly anticommutative R algebra. Let M be any R module and let $0 \to K \to F \to M \to 0$ be exact with F free. Then $\Lambda(M) = \Lambda(F)/\Lambda(F)i(K)$. To show this we must show that if G is a strictly anticommutative R algebra and $f\colon M \to G_1$ is a homomorphism of R modules, then there is an algebra map $\overline{f}\colon \Lambda(F)/\Lambda(F)i(K) \to G$. Since $\Lambda(F)$ is the exterior algebra, there is a map $\Lambda(F) \to G$ coming from the composite map of R modules $F \to M \to G_1$ such that the diagram

$$\begin{array}{ccc} F & \longrightarrow & M \\ \downarrow & & \downarrow \\ \Lambda(F)_1 & \longrightarrow & \Lambda(M)_1 \end{array} \quad \text{commutes.}$$

Hence K maps to 0 in $\Lambda(M)_1$ under the map $\Lambda(F) \to G_1$. Therefore, the image of $\Lambda(F)i(K)$ is 0 in G. Thus, the map $\Lambda(F) \to G$ factors through $\Lambda(M)$. Therefore, $\Lambda(M)$ is the exterior algebra as claimed.

<u>Remark</u>. Keeping the notation above, we see that $\Lambda(M)_0 = R$ and $\Lambda(M)_1 = M$. This is clear for free modules. If M is not free, then $i(K)\Lambda(F)$ is 0 in degree 0 and K in degree 1. Hence, the remark is true in this case also.

<u>Proposition 8.1</u>. a) Let $f\colon R \to R'$ be a homomorphism of commutative rings and let M be an R module. Then $\Lambda_{R'}(R' \otimes_R M)$ is isomorphic to $R' \otimes_R \Lambda_R(M)$ where the subscripts indicate which ring we are taking the exterior algebra over.

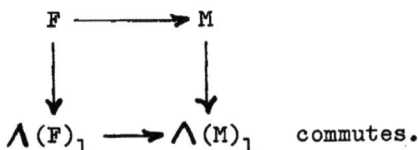

b) Let M and N be R modules. Then $\Lambda(M \oplus N)$ is isomorphic to $\Lambda(M) \otimes_R \Lambda(N)$ where the product in the tensor product is given $(m \otimes n)(m' \otimes n') = (-1)^{rs}(mm' \otimes nn')$ where degree $(m') = r$ and degree $(n) = s$.

Proof. a) follows from the definition of Λ as an adjoint. Let G be strictly anticommutative R' algebra. Then $\mathrm{Hom}_{R'\text{-alg}}(\Lambda_{R'}(R' \otimes_R M), G)$ is isomorphic to $\mathrm{Hom}_{R'\mathrm{mod}}(R' \otimes_R M, G)$ is isomorphic to $\mathrm{Hom}_{R\text{-mod}}(M, G_1)$ is isomorphic to $\mathrm{Hom}_{R\ \mathrm{alg}}(\Lambda_R(M), G)$ is isomorphic to $\mathrm{Hom}_{R'\mathrm{alg}}(R' \otimes_R \Lambda_R(M), G)$.

b) Since Λ is an adjoint functor, it is enough to show that tensor product is direct sum in the category of strictly anticommutative graded R algebras. It is clear that the tensor product of two strictly anticommutative R algebras is a strictly anticommutative R algebra. The proof that tensor product is direct sum in the category is exactly the same as in the category of commutative rings.

Done.

Proposition 8.2. Let R be a commutative ring and P be a finitely generated projective R module of constant rank $(r_P(\underline{p}) \equiv n)$. Then $\Lambda(P)_i = 0$ for $i > n$ = rank of P, $\Lambda(P)_i$ is a finitely generated projective for all n, and $\Lambda(P)_n$ has rank 1.

Proof. Case 1). P is free on n generators, $e_1, \ldots, e_n$, then $\Lambda(P)_i$ is free on the generators $e_{\alpha_1}, \ldots, {\alpha_i}$ where $\alpha_1 < \ldots < \alpha_i$. The proposition is clear in this case.

Case 2). P any finitely generated projective of rank n. By proposition 8.1 a) $\Lambda(P)_{\underline{p}} = \Lambda(P_{\underline{p}})$. But $P_{\underline{p}}$ is free for all prime ideal $\underline{p}$. Therefore $\Lambda(P)_{n_0} = 0$ when $n_0 > 0$ since if M is an R module such that $M_{\underline{p}} = 0$ for all prime ideals $\underline{p}$ then $M = 0$. $\Lambda(P)_{i_{\underline{p}}} = \Lambda(P_{\underline{p}})_i$ is a finitely generated free $R_{\underline{p}}$ module and $\Lambda(P)_i$ is clearly finitely generated. Therefore, $\Lambda(P)_i$ is a finitely generated projective R module. Similarly $\Lambda(P)_n$ has rank 1. Done.

Let P be a projective R module of constant rank n. We define $\det_R(P) = \Lambda(P)_n$. This compares with the usual determinant in the following way. Let $f: P \longrightarrow P$ be a homomorphism of R modules. Then $Q = \Lambda(P)_n$ is finitely generated of rank 1. Hence $\Lambda(f) \in \mathrm{Hom}_R(Q, Q)$ which is isomorphic to R. Hence $\Lambda(f) \in R$. If P were free, then this would be the usual determinant of f. In fact, this is the definition of determinant used by Bourbaki.

<u>Proposition 8.3</u>. a) Let $f: R \longrightarrow R'$ be a homomorphism of commutative rings and let P be a finitely generated projective R module of constant rank. Then $R' \otimes_R \det_R(P)$ is isomorphic to $\det_{R'}(R' \otimes_R P)$.

b) If P and Q are finitely generated projective R modules of constant rank m and n respectively, then $\det_R(P \oplus Q) = \det_R(P) \otimes_R \det_R(Q)$.

<u>Proof</u>. a) is immediate from Proposition 8.1.a).

b) By proposition 8.1.b) $\Lambda(P \oplus Q) = \Lambda(P) \otimes_R \Lambda(Q)$. Hence $\Lambda(P \oplus Q)_{m+n} = \coprod_{i+j=m+n} \Lambda(P)_i \otimes_R \Lambda(Q)_j$. But if $i > m$ or $j > n$ $\Lambda(P)_i \otimes_R \Lambda(Q)_j = 0$. Hence $\Lambda(P \oplus Q)_{m+n} = \Lambda(P)_m \otimes_R \Lambda(Q)_n$. $P \oplus Q$ clearly has constant rank $n + m$. Therefore, $\det_R(P \oplus Q) = \det(P) \otimes_R \det(Q)$.

Done.

<u>Corollary 8.4</u>. If R is a commutative ring such that every finitely generated projective has constant rank, then det factors through $K_0(R)$ and the factored map is a homomorphism of abelian

groups. The composite map $Pic(R) \xrightarrow{i} K_0(R) \xrightarrow{det} Pic(R)$ is the identity.

Proof. The first part is clear by Proposition 8.3. If P is a finitely generated rank 1 projective then $det_R(P) = \Lambda(P)_1 = P$ since for any module M, $\Lambda(M)_1 = M$. Done.

Before we can generalize this result to any commutative ring we must prove a few general theorems.

Let $\underline{C}$ denote the category of sets, monoids, groups, rings, or abelian groups. (That is, $\underline{C}$ is a category that one can use to build sheaves.)

Theorem 8.5. Let F be a covariant functor from the category of commutative rings to $\underline{C}$. Then there is a functor $F^{\#}$ and a natural transformation $\eta: F \longrightarrow F^{\#}$ such that $F^{\#}$ preserves finite products and η is universal for such natural transformations.

Remark. This theorem is analogous to constructing the sheaf associated to a presheaf over a Grothendieck topology. In fact it is precisely this if we call a "covering" of spec (R) a decomposition into disjoint open sets.

Proof. Let R be a commutative ring and let D_R be the set of decompositions of R into a finite direct product of rings. D_R is also the set of all decompositions of 1 into a sum of orthogonal idempotents. Let d and $d' \in D_R$. We say $d \geqslant d'$ if the decomposition of R given by d' is a refinement of the one given by d. That is if $1 = \sum_I e_i$ is the decomposition given by d and $1 = \sum_J e_j'$ is the decomposition given by d' then $e_i = \sum_{j \in s_i} e'_j$ where s_i is a subset of J for each $i \in I$.

If $1 = \sum_I e_i$ and $1 = \sum_J e'_j$ are two decompositions then $1 = \sum_{I \times J} e_i e'_j$ is a decomposition which is finer than both of them. Hence D_R is a directed set.

Applying F to each element of D_R and using the maps between refinements we get a directed system in $\underline{C}$ for each ring. We define $F^{\#}(R) = \varinjlim$ over the system obtained from D_R. η_R: $F(R) \longrightarrow F^{\#}(R)$ is obtained from the identity map of F(R) to F(R) where R = R is considered as the initial element of D_R. It is clear that $F^{\#}$ is a functor and that η_R is natural with respect to R and F. If F preserves finite products, then η is an isomorphism. For, if $R = R_1 x \ldots x R_n$, then F(R) is isomorphic to $F(R_1) x \ldots x F(R_n)$. Hence η is an isomorphism for every $d \in D_R$.

$F^{\#}$ preserves finite products since directed $\varinjlim$ commutes with finite products in $\underline{C}$.

Since $F^{\#}$ preserves finite products and directed $\varinjlim$ commutes with finite products in $\underline{C}$ $\eta^{\#}$: $F^{\#} \longrightarrow F^{\#\#}$ is an iso.

Let v: $F \longrightarrow G$ be a natural transformation of functors where G preserves finite products. Then the diagram

$$\begin{array}{ccc} F & \xrightarrow{v} & G \\ \eta \downarrow & & \downarrow \eta \\ F^{\#} & \xrightarrow{v^{\#}} & G^{\#} \end{array}$$

commutes.

and η: $G \longrightarrow G^{\#}$ is an iso. Hence $\eta^{-1} v^{\#}$: $F^{\#} \longrightarrow G$ is a natural transformation of functors.

Let $\underline{a}$ and $\underline{b}$: $F^{\#} \longrightarrow G$ be two natural transformations making the diagram

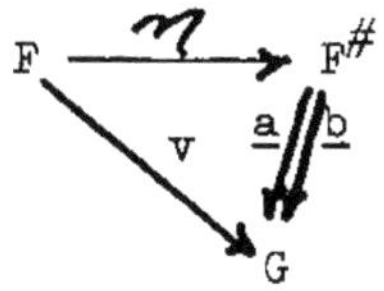

commute.

We want to show that $\underline{a} = \underline{b}$. Since $\eta^{\#}$ is an iso on $F^{\#}$ and $G^{\#}$ it is enough to examine the case

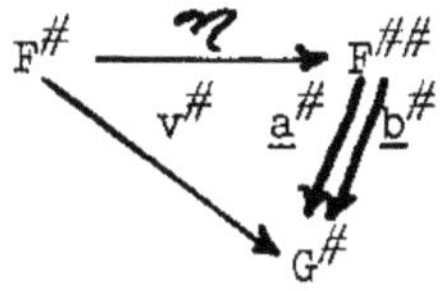

and prove that $\underline{a}^{\#} = \underline{b}^{\#}$. But $\eta^{\#}: F^{\#} \longrightarrow F^{\#\#}$ is an iso. Hence it is enough to show that $\underline{a}^{\#}\eta^{\#} = \underline{b}^{\#}\eta^{\#}$. But the diagram commutes. Hence $\underline{a}^{\#}\eta^{\#} = v^{\#}\eta^{\#}$ and $\underline{a} = \underline{b}$ as desired. Therefore there is a unique transformation $F^{\#} \longrightarrow G$ whenever we are given $v: F \longrightarrow G$ and G preserves finite products. Done.

Corollary 8.6. Let F and G be covariant functors from the category of commutative rings to $\underline{C}$ and $v: F \longrightarrow G$ be a natural transformation. If G preserves finite direct products then the transformation $\underline{a}: F^{\#} \longrightarrow G$ is a natural iso if and only if 1) If x and $y \in F(R)$ and $v(x) = v(y)$ in $G(R)$ then there exists a decomposition $R = R_1 x \ldots x R_n$ such that x and y go to the same thing in each $F(R_i)$, and 2) If $x \in G(R)$, there exists $R = R_1 x \ldots x R_y$ such that the image of x in each $G(R_i)$ comes from $F(R_i)$ by v.

Proof. This is immediate from the construction in the proof of theorem 8.5. and the usual characterization of what $\varinjlim$ are in $\underline{C}$. Done.

K_0, H, and Pic are all covariant functors preserving finite direct products. Let P(R) be the set of isomorphism classes of finitely generated R modules. P(R) is a monoid where $\oplus$ gives the addition.

Consider the constant functor $\underline{Z}$: Category of commutative rings to $\underline{Ab}$ where all objects go to Z and all maps to 1_Z. We claim that $\underline{Z}^{\#} = H$. We map $\underline{Z} \xrightarrow{f} H$ by $\underline{Z}(R) = Z \xrightarrow{f_R} H(R)$ where $f_R(n)$ is the constant function n. We check the condition of corollary 8.6. Condition 1) is obvious. For 2) let $f \in H(R)$. Then $\text{spec}(R) = U_1 \cup \ldots \cup U_n$ where $f^{-1}(n_i) = U_i$. This gives $R = R_1 x \ldots x R_n$. $H(R) = H(R_1) x \ldots x H(R_n)$. f goes to the constant function n_i on R_i and hence f comes from $Z(R_i)$ via f.

We defined g: $\underline{Z} \longrightarrow K_0$ by $g_R(n) = [R^n]$. Since $\underline{Z}^{\#} = H$ we get a natural transformation $H \longrightarrow K_0$. This is clearly the same natural transformation as the earlier one. Let PC(R) be the set of isomorphism classes of finitely generated projective R modules of constant rank. PC(R) is a monoid with direct sum as addition. Let $PC \longrightarrow P$ be the obvious inclusion. Let $Q \in P(R)$, then r_Q breaks spec(R) into a disjoint union of open sets $U_1, \ldots, U_n$ where $r_Q(\underline{q}) = n_i$ for all $q \in U_i$. Let $R = R_1 x \ldots x R_n$ be the corresponding decomposition of R. We claim that $PC^{\#}$ is isomorphic to P. Since PC is included in P, condition 1) of corollary 8.6. is satisfied. Q is isomorphic to $Q_1 \oplus \ldots \oplus Q_n$ where $Q_i = R_i \otimes_R Q$ shows that condition 2) is satisfied. Let $\underline{r}_R$: $PC(R) \longrightarrow Z(R)$ be the rank map. Then this gives a natural transformation $\underline{r}^{\#}$: $PC^{\#} \longrightarrow Z^{\#}$. This is the same as giving a natural transformation $\underline{r}^{\#}$: $P \longrightarrow H$. These maps yield maps of monoids $\underline{r}_R^{\#}$: $P(R) \longrightarrow H(R)$. Therefore the map $\underline{r}^{\#}$ factors through K_0. The map p: $K_0 \longrightarrow H$ is the generalized rank map.

Now we have the natural transformation of functors det: PC $\longrightarrow$ Pic. Pic preserves products. Hence there is a unique natural transformation of functors $PC^{\#}$ = P $\longrightarrow$ Pic. But P(R) $\longrightarrow$ Pic(R) is a map of monoids. Hence P $\longrightarrow$ Pic factors through K_0. Call the factored map $\underline{\text{det}}$. Then $\underline{\text{det}}$(Q) = det(Q) if Q has constant rank. If Q has constant rank 1, $\underline{\text{det}}$(Q) = det(Q) = Q. Hence if i: Pic $\longrightarrow K_0$ is given by i([P]) = [P] and if i': Pic $\longrightarrow K_0$ is given by i'([P]) = [P] - [R] then $\text{det}\cdot i = 1_{\text{Pic}}$ and $\text{det}\cdot i' = 1_{\text{Pic}}$. The last since [R] is the identity in Pic(R). This completes the generalization of corollary 8.4. and accomplishes the goal of the chapter.

Chapter 9. Basic Topological Remarks.

Definition. Let X be a non empty topological space. Then X is *irreducible* if X = A $\cup$ B with A and B closed implies A = X or B = X.

Examples. 1) A point. If X is hausdorf , then X is irreducible if and only if X is a point. If x and y are distinct points of a hausdorf space X, then there exist disjoint open sets U and V such that $x \in U$ and $y \in V$. Then X = (X - U) $\cup$ (X - V) both proper closed sets.

2) If X is an infinite set with the finitary topology, then X is irreducible. These examples can be thought of as algebraic curves with the Zariski topology.

Proposition 9.1. Let A be a non empty subspace of a topological space X. Then A is irreducible if and only if its closure $\bar{A}$ is.

Proof. $\Longrightarrow$ Let $\bar{A}$ = B $\cup$ C with B and C closed. Then $A = \bar{A} \cap A = (B \cap A) \cup (C \cap A)$. Hence B or C is A and $\bar{B}$ or $\bar{C}$ is $\bar{A}$.

$\Leftarrow$ Let $A = B \cup C$ with B and C closed in A. Then $\bar{A} = \bar{B} \cup \bar{C}$. $\bar{A}$ irreducible implies $\bar{B} = \bar{A}$ or $\bar{C} = \bar{A}$. Say $\bar{B} = \bar{A}$. B closed in A means $B = \bar{B} \cap A$. Therefore, $B = A$. Done.

Corollary 9.2. Let X be any topological space and let $x \in X$. Then $\bar{x}$ is irreducible.

Definition. Let A be a closed set of a topological space X and let $a \in A$. Then a is a generic point of A if $\bar{a} = A$.

Definition. Let X be a topological space. Then the combinatorial dimension of X, dim X, is the supremum of the n such that there is a chain of non empty closed irreducible subsets of X, $\emptyset \neq A_0 < A_1 < \ldots < A_n$.

Examples. dim (point) = 0. dim (example 2 above) = 1. dim (any hausdorff space) = 0. We can choose dim $\emptyset = -1$ since the definition does not apply if $X = \emptyset$.

Proposition 9.3. Let A be a non empty subspace of a topological space X. Then dim $A \leqslant$ dim X.

Proof. Let $A_0 < A_1 < \ldots . < A_n$ be a chain of non empty relatively closed irreducible subsets of A. Then $\bar{A}_0 \leqslant \bar{A}_1 \leqslant \ldots \leqslant \bar{A}_n$ where closure is taken in X. $A_i = \overline{A_i} \cap A$. Hence if $\overline{A_i} = \overline{A_{i+1}}$ then $A_i = A_{i+1}$. Therefore, $\bar{A}_0 < \ldots < \bar{A}_n$ is a chain of non empty relatively closed subsets of X. By proposition 9.1. $\bar{A}_i$ are all irreducible.

Done.

Corollary 9.4. dim (m-spec (R)) $\leqslant$ dim (spec (R)).

Definition. Let A be a non empty subset of spec(R) where R is a commutative ring. I(A) is the largest ideal, $\underline{a}$, of R such that $F(\underline{a}) \supset A$.

Remarks. I(A) always exists for let $\underline{a}_i$ be all ideals of R such that $F(\underline{a}_i) \supset A$. Let $\underline{a} = \sum \underline{a}_i$. Then $F(\underline{a}) = \bigcap F(a_i) \supset A$ and $\underline{a} = I(A)$. It is clear that $I(A) = \bigcap_{\underline{p} \in A} \underline{p}$.

Proposition 9.5. Let A be a nonempty subset of spec(R). Then A is irreducible if and only if I(A) is a prime ideal.

Proof. $\Longrightarrow$ Let $I(A) = \underline{a}$. Let $\underline{b}$ and $\underline{c}$ be ideals of R such that $\underline{bc} \subset \underline{a}$. We must show $\underline{b} \subset \underline{a}$ or $\underline{c} \subset \underline{a}$.

$F(\underline{bc}) \supset F(\underline{a}) = A$. But $F(\underline{bc}) = F(\underline{b}) \cup F(\underline{c})$. A is irreducible. Therefore, either $F(\underline{b}) \supset F(\underline{a})$ or $F(\underline{c}) \supset F(\underline{a})$. Hence, either $\underline{b} \subset \underline{a}$ or $\underline{c} \subset \underline{a}$. Thus, $\underline{a}$ is a prime ideal.

$\Longleftarrow$ By proposition 9.1. we can assume that A is closed. Since $FI(A) = \bar{A}$, we have that $FI(A) = A$. Let $I(A) = \underline{a}$ be a prime ideal. Suppose $A = B \cup C$ where B and C are closed in A. Then B and C are closed in spec(R). Therefore, $B = F(\underline{b})$ and $C = F(\underline{c})$ where $\underline{b}$ and $\underline{c}$ are ideals of R. $B \cup C = A$ implies $F(\underline{b}) \cup F(\underline{c}) = F(\underline{bc}) = F(\underline{a})$. Therefore, $\underline{a} \supset \underline{bc}$. But $\underline{a}$ is prime. Thus, $\underline{b} \subset \underline{a}$ or $\underline{c} \subset \underline{a}$. In the first case, $B = F(b) \supset A$; in the second case $C = F(\underline{c}) \supset A$. Therefore A is irreducible. Done.

Corollary 9.6. There is a one-one correspondence between prime ideals of R and irreducible closed sets of spec(R) given by $\underline{p}$, a prime ideal, goes to $F(\underline{p})$, and A, a closed irreducible subset, goes to I(A).

Corollary 9.7. Each closed irreducible subset of spec(R) has a unique generic point.

Remarks. The property in corollary 9.7 is a local property. That is if X is a topological space with that property, then every open subset has that property; and if $X = \bigcup U_i$ with the U_i open and if each U_i has that property, then X has that property. Thus corollary 9.7. remains true for schemes.

Corollary 9.7. is false for m-spec(R) since any point is closed but there are irreducible closed sets which contain more than one point in general.

Definition. Let X be a topological space. Then X is a noetherian space if X has the descending chain condition on closed sets.

Examples. Let R be a noetherian ring. Then spec(R) is a noetherian space. Let $A_1 \supset A_2 \supset \ldots . \supset A_n \supset \ldots$ be a descending chain of closed sets. Then $I(A_1) \subset I(A_2) \subset \ldots \subset I(A_n) \subset \ldots$ is an ascending chain of ideals. Hence there is an n such that if $m > n$ $I(A_n) = I(A_m)$. But the A_i's are all closed. Hence $FI(A_i) = A_i$. Therefore, the chain of A_i's stops at n also.

Remark. The converse of the above is false. Let R be a discrete rank 2 valuation ring. Then spec(R) has 3 points and, hence, spec(R) is a noetherian space. R is not a noetherian ring.

Proposition 9.8. If X is a noetherian space, then all subspaces of X are noetherian.

Proof. Let Y be a subspace of X and let $A_1 \supset \ldots \supset A_n \supset \ldots$ be a chain of closed subsets of Y. Then the chain $\bar{A}_1 \supset \ldots \supset \bar{A}_n \supset \ldots$ of closed subsets of X stops for some n. But $A_i = \bar{A}_i \cap Y$. Therefore the chain $A_1 \supset \ldots . \supset A_n$ stops for the same n.

Corollary 9.9. m-spec(R) is noetherian if R is noetherian.

Proposition 9.10. If X is a noetherian space, then $X = X_1 \cup \ldots \cup X_n$ where n is finite and the X_i's are closed and irreducible.

Proof. Consider the closed sets, A, of X such that the proposition is false for A. If there are any, there is a minimal one, A.

Case 1. A is irreducible. This is an immediate contradiction.

Case 2. A is not irreducible. Then $A = B \cup C$ where B and C are proper closed subsets of A. Then the result is true for B and C. Therefore, the result is true for A. This is a contradiction.

Therefore, the result is true for every closed set of X. In particular, it is true for X. Done.

If $A \subset X$ is irreducible and X is noetherian, then $A = A \cap X = (A \cap X_1) \cup \dots \cup (A \cap X_n)$. Hence $A \subset X_i$ some i. If we write $X = X_1 \cup \dots \cup X_n$ such that $X_i \not\subset \bigcup_{i \neq j} X_j$ then the X_i are the maximal irreducible closed subsets of X. Hence the decomposition is unique if we require that the X_i be maximal irreducible subsets. These subsets are called the irreducible components of X. If X is not noetherian, we can use Zorn's lemma on an increasing chain of irreducible closed sets to prove every irreducible is contained in a maximal one and that X = U maximal irreducible closed sets.

Definition. Let X be a topological space and let $A \subset B$ be closed subsets of X. Then if A is irreducible $\underline{\operatorname{codim}_B A}$ = sup of all n such that there exists a chain

$$A = A_0 < A_1 < \dots < A_n \subset B \text{ with } A_i$$

closed and irreducible. If A is a closed set then $\underline{\operatorname{codim}_B A} = \min \operatorname{codim}_B A'$ where A' runs over the irreducible components of A.

Proposition 9.11. Let X be a topological space and let A, B, and C be closed subsets of X.

1) $\operatorname{codim}_B A \leqslant \dim B$.

2) $\operatorname{codim}_B A = \min \operatorname{codim}_B A'$ where A' is any closed irreducible subset of A.

3) If $A \subset B \subset C$, then $\operatorname{codim}_C A \geqslant \operatorname{codim}_C B$.

4) If $A \cup B \subset C$, then $\operatorname{codim}_C(A \cup B) = \min(\operatorname{codim}_C A, \operatorname{codim}_C B)$.

5) If $A \subset B \subset C$, then $\operatorname{codim}_C A \geqslant \operatorname{codim}_C B + \operatorname{codim}_B A$.

Proof. 1) is obvious.

2) $\geqslant$ is clear. Let $A' \subset A$ be a closed irreducible subset. Then there exists A" with $A' \subset A'' \subset A$ where A" is a maximal closed irreducible set contained in A. Then $\operatorname{codim}_B A' \geqslant \operatorname{codim}_B A''$. Hence the min taken over the irreducible components of A $\leqslant$ min taken over all irreducible closed subsets of A. Hence equality.

3) This is clear from 2) since any closed irreducible set of A is a closed irreducible set of B.

4) D is a closed irreducible set of $A \cup B$ if and only if D is a closed irreducible set of A or D is a closed irreducible set of B. Apply 2).

5) Let $A' \subset A$ be a closed irreducible set of A. Let

$A' = A_0 < \dots < A_n \subset B$ and

$A_n = B_0 < B_1 < \dots < B_m \subset C$ be two chains of closed irreducible sets. Then

$A' = A_0 < \dots < A_n = B_0 < B_1 < \dots < B_m \subset C$ is a chain with length $n + m$. Let p be the $\operatorname{codim}_C B$ if finite and be arbitrarily large if $\operatorname{codim}_C B$ is infinite. Let $q = \operatorname{codim}_B A$ if finite and arbitrarily large if $\operatorname{codim}_B A$ is infinite. Then $\operatorname{codim}_C A \geqslant p + q$ is clear.

Done.

Chapter 10. Chain Complexes and the Nilpotence of $\widetilde{K_0(R)}$

The goal of this chapter is to prove that if R is a noetherian commutative ring with dim(m-spec(R)) = d then $\widetilde{K_0(R)}^{d+1} = 0$.

Definition. Let R be a ring. A finite projective chain complex over R is:

a) a sequence of R modules and homomorphisms

$$\longrightarrow C_{n+1} \xrightarrow{d_{n+1}} C_n \xrightarrow{d_n} C_{n-1} \longrightarrow \text{ with } d_n d_{n+1} = 0, \text{ and}$$

b) C_n are all finitely generated projective R modules with all but a finite number of the $C_n = 0$.

If we are given an exact functor T: $\underline{A} \longrightarrow \underline{B}$ between abelian categories then T commutes with homology. That is $H_n(TC)$ is isomorphic to $TH_n(C)$ where C is any chain complex. In particular the functor $M \rightsquigarrow M_m$, localizing at maximal ideals, is exact. Hence, $H_n(C)_m = H_n(C_m)$.

Definition. Let $\underline{C}$ be a finite projective chain complex over a ring R. Then $\mathrm{Supp}(H(\underline{C})) = \{m | H_n(\underline{C})_m \neq 0$ for at least one n and m is a maximal ideal of R$\}$.

Remarks. If R is noetherian then C_n finitely generated implies $H_n(\underline{C})$ is finitely generated. Hence $\mathrm{Supp}\, H(\underline{C}) = \bigcup_n \mathrm{Supp}\, H_n(\underline{C})$ and is a closed set of m-spec(R). If m is a maximal ideal then $m \notin \mathrm{Supp}\, H(\underline{C})$ if and only if the complex C_m is an exact sequence.

Definition. Let $\underline{C}$ be a finite projective chain complex over R. We define $[\underline{C}] \in K_0(R)$ by $[\underline{C}] = \sum (-1)^n [C_n]$.

Definition. Let $X \in K_0(R)$. Then $X \in F^D K_0(R)$ if for every closed set $B \subset$ m-spec(R) there is a finite projective chain complex $\underline{C}$ such that

1) $[\underline{C}] = X$, and

2) $\operatorname{codim}_B[B \cap \operatorname{Supp} H(\underline{C})] \geq p$.

Lemma 10.1. Let $\underline{C}$ be a finite projective chain complex. Then the following are equivalent:

1) $H(\underline{C}) = 0$,

2) $\underline{C}$ is a finite direct sum of complexes of the form $\to 0 \to \dots \to 0 \to P \xrightarrow{f} P' \to 0 \to \dots$ where f is an iso and P is projective, and

3) $\underline{C}$ is contractible. That is there exist module homomorphisms $S_n: C_n \to C_{n+1}$ such that $dS + Sd = 1$.

Proof. 1) $\Longrightarrow$ 2) $\underline{C}$ is $0 \to \dots \to C_n \to \dots \to C_m \to 0 \to \dots$ where $C_n, C_{n-1}, \dots, C_m$ are all the non zero terms. We proceed by induction on $n - m$. If $n - m = 0$ then $\underline{C} = G \to C_n \to 0$ and $0 = H(C_n) = C_n$. Hence can assume $n - m > 0$. We have

$$\to \dots \to C_{m+1} \to C_m \to 0 \to \dots$$

$C_{m+1} \to C_m$ is an epi since $H(\underline{C}) = 0$. C_m is projective so there exists a splitting $C_{m+1} = C''_{m+1} \oplus C'_{m+1}$ where $C'_{m+1} = \ker d_{m+1}$. Then $\underline{C}$ is the sum of the complexes

$$0 \to C_n \to \dots \to C_{m+2} \to C'_{m+1} \to 0 \text{ and}$$
$$0 \to C''_{m+1} \to C_m \to 0.$$

The proof is finished by induction.

2) $\Longrightarrow$ 3) If $\underline{C}$ and $\underline{C}$ are contractible chain complexes then so is $\underline{C} \oplus \underline{C}'$ and $0 \to P \xrightarrow{f} P' \to 0$ where f is an iso is obviously contractible.

3) $\Longrightarrow$ 1) Let $dx = 0$. Then $x = (dS + Sd)x = dSx$. Hence every cycle is a boundary and $H(\underline{C}) = 0$. Done.

Recall that for a commutative ring R, $K_0(R)$ can be made into a commutative ring with unit by defining $[P][Q] = [P \otimes_R Q]$.

Proposition 10.2. Let R be a commutative noetherian ring. Then

1) $F^0 = K_0 \supset F^1 \supset \dots$

2) F^p is an additive subgroup of $K_0(R)$.

3) $F^p F^q \subset F^{p+q}$.

4) If dim $\widetilde{\text{m-spec}}(R) = d$, then $F^{d+1} = 0$.

5) $F^1 \subset \widetilde{K_0(R)}$

6) $F^1 \supset \widetilde{K_0(R)}$ ie $F^1 = \widetilde{K_0(R)}$.

Remark. 1)-5) are quite formal and are even true for noetherian preschemes under the proper generalization of the definition of F^i. 6) needs R to be a ring.

Proof. 1) Since codim is always ≥ 0, if we can represent every $x \in K_0(R)$ as $[\underline{C}]$, then $F^0 K_0(R) = K_0(R)$. But $x \in K_0(R)$, then $x = [P] - [Q]$ where P and Q are finitely generated projective R modules. The complex $0 \longrightarrow Q \xrightarrow{f} P \longrightarrow 0$ where Q is in C_1's place, P is in C_0's place and $f = 0$ clearly represents x. $F^i \supset F^{i+1}$ is clear.

2) First we show if $x \in F^p$ then $-x \in F^p$. Given B we find $\underline{C}$, a finite projective chain complex, such that $x = [\underline{C}]$ and Supp $H(C) \cap B$ has codim $\geq p$ in B. Define a finite projective chain complex C' by $C'_n = C_{n+1}$ and $d'_n \colon C'_n \longrightarrow C'_{n-1}$ is d_{n+1}. Then $\underline{C}'$ represents $-x$. Supp $\underline{C}$ = Supp $\underline{C}'$. Therefore $-x \in F^p$.

Let x and $y \in F^p$ we want to show that $x + y \in F^p$. Suppose we are given a closed set B. Then $x = [\underline{C}]$ and $y = [\underline{C}']$ where $\underline{C}$ and $\underline{C}$ are finite projective chain complexes satisfying the codim requirement for B. The complex $\underline{C} \oplus \underline{C}'$ represents $x + y$ since

$[\underline{C} \oplus \underline{C}'] = \sum(-1)^n[C_n \oplus C_n'] = \sum(-1)^n C_n + \sum(-1)^n C_n' = x + y.$

$H(\underline{C} \oplus \underline{C}') = H(\underline{C}) \oplus H(\underline{C}')$. Supp $H(C \oplus C') \subset$ Supp $H(C) \cup$ Supp $H(C')$. Therefore, $\mathrm{codim}_B(B \cap \mathrm{Supp}\, H(\underline{C} \oplus C') \geq \min(\mathrm{codim}_B(B \cap \mathrm{Supp}\, H(\underline{C}))$, $\mathrm{codim}_B(B \cap \mathrm{Supp}\, H(\underline{C}'))$ by proposition 9.11.4. Hence $\underline{C} \oplus \underline{C}'$ satisfies the codim requirement for B and $x + y \in F^p$.

4) Let $x \in F^{d+1}$. Let B = m-spec(R). Then $x = [\underline{C}]$ and $\mathrm{codim}_B B \cap \mathrm{Supp}\, H(\underline{C}) \geq d + 1$. Therefore Supp $H(\underline{C}) \cap$ m-spec(R) has codim d + 1 in m-spec(R). That is Supp $H(\underline{C}) = \emptyset$. Therefore, for all maximal ideals m, $H(\underline{C})_m = 0$. Therefore, $H(C) = 0$ and $\underline{C}$ is an exact sequence. Thus, $x = 0$.

3) $F^pF^q \subset F^{p+q}$. Let $x \in F^p$ be represented by $\underline{C}$ for the closed set B and $y \in F^q$ be represented by $\underline{C}'$ for B. Then $(C \otimes C')$ represents xy where $(C \otimes C')_n = \coprod_{i+j=n} C_i \otimes C_j$. Let t: $\underline{C} \to \underline{C}$ by t is +1 on $\underline{C}_n$ if n is even and t is -1 on C_n if n is odd. Then $d_{C \otimes C'} = d_C \otimes 1 + t \otimes d_C$. $d_{C \otimes C'} d_{C \otimes C'} = 0$ for $d_{C \otimes C'} d_{C \otimes C'} = d_C^2 \otimes 1 + t^2 \otimes d_{C'}^2 + d_C t \otimes d_{C'} + t d_C \otimes d_{C'}$. But $d_C^2 \otimes 1 = 0$ and $t^2 \otimes d_{C'}^2 = 0$. Hence $d_{C \otimes C'} d_{C \otimes C'} = d_C t \otimes d_{C'} + t d_C \otimes d_{C'} = 0$.

$$
\begin{aligned}
[\underline{C} \otimes \underline{C}'] &= \sum(-1)^n[C \otimes C']_n = \sum_{i,j}(-1)^{i+j}[C_i \otimes C_j'] \\
&= \sum(-1)^{i+j}[C_i][C_j'] \\
&= (\sum(-1)^i[C_i])(\sum(-1)^i[C_j']).
\end{aligned}
$$

Claim: If $\underline{C}$ and $\underline{C}'$ are finite projective chain complexes and $H(C) = 0$ then $H(\underline{C} \otimes \underline{C}') = 0$. $\underline{C}$ is contractible by lemma 10.1. We have $S: C \longrightarrow C$ with $dS + Sd = 1$. Then $S \otimes 1$ is a contracting map for $\underline{C} \otimes \underline{C}'$. Since

$$(S \otimes 1)(d \otimes 1 + t \otimes d) + (d \otimes 1 + t \otimes d)(S \otimes 1) =$$
$$= (Sd + dS) \otimes 1 + (tS + St) \otimes 1$$
$$= \quad 1 \otimes 1 \quad + \quad 0 \otimes 1$$
$$= \quad 1.$$

Given a closed set $B \subset$ m-spec(R) we find $x = [C]$ where $\operatorname{codim}_B(B \cap \operatorname{Supp} H(\underline{C})) \geq p$. Let $A = B \cap \operatorname{Supp} H(\underline{C})$ (a closed set). Find $y = [\underline{C}']$ where $\operatorname{codim}_A(A \cap \operatorname{Supp} H(\underline{C}')) \geq q$. Let $D = A \cap \operatorname{Supp} H(\underline{C}')$.

$\operatorname{Supp} H(\underline{C} \otimes \underline{C}') \subset \operatorname{Supp} H(\underline{C}) \cap \operatorname{Supp} H(\underline{C}')$. For $\underline{p} \in \operatorname{Supp} H(C \otimes C')$ if and only if $H(C \otimes C')_{\underline{p}} \neq 0$, if and only if $H((\underline{C} \otimes \underline{C}')\underline{p}) \neq 0$. But $H((\underline{C} \otimes \underline{C})_{\underline{p}}) = H(\underline{C}_{\underline{p}} \otimes_{R_{\underline{p}}} \underline{C}_{\underline{p}})$. $\underline{p} \in \operatorname{Supp} H(C)$ if and only if $H(C_{\underline{p}}) \neq 0$ and the claim above.

$$\operatorname{codim}_B(B \cap \operatorname{Supp} H(\underline{C} \otimes \underline{C}')) \geq \operatorname{codim}_B D \geq \operatorname{codim}_B A + \operatorname{codim}_A D \geq p + q$$

by proposition 9.11.5.

5) $F^1 \subset \widetilde{K}_0$. First we prove a lemma.

<u>Lemma 10.3</u>. $x \in F^1K_0(R)$ if and only if for each finite set $S \subset$ m-spec(R) there exists a finite projective chain complex $\underline{C}$ such that $x = [\underline{C}]$ and $H(\underline{C}_m) = 0$ for all $m \in S$.

<u>Proof</u>. Let $x = [\underline{C}]$ and $A = S \cap \operatorname{Supp} H(\underline{C})$. If $X \in F^1$ this is clear from the definition of F^1 (with $B = S$) i.e., with $\operatorname{codim}_S A \geq 1$. Then $A = \emptyset$, so $H(\underline{C}_m) = 0$ for all $m \in S$.

Conversely, if x has the stated property and B is any closed set, let $B = B_1 \cup \ldots \cup B_n$ be the decomposition into irreducible components. Choose $b_i \in B_i - \bigcup_{j \neq i} B_j$ and let $S = \{b_1, \ldots, b_n\}$.

If $x = [C]$, $H(\underline{C})_m = 0$ for $m \in S$, then no $B_i \subset B_n \cap \operatorname{supp} H(\underline{C})$ so this set has codim ≥ 1 in B.

Next we observe that $\widetilde{K_0(R)} = \bigcap_{m \text{ maximal}} \ker r_m$ where

$r_m \colon K_0(R) \longrightarrow K_0(R_m)$ by $[P] \longrightarrow [P_m]$. $\widetilde{K_0(R)} = \bigcap_{\underline{p} \text{ prime}} \ker r_{\underline{p}}$

but if $\underline{p}$ is a prime ideal it is contained in a maximal ideal m, and we localize at $\underline{p}$ by localizing at m and then localizing R_m at $\underline{p}_m$. Hence, $\ker r_{\underline{p}} \supset \ker r_m$. Let $x \in F^1K_0(R)$. We must show that x goes to 0 in each $K_0(R_m)$ where m is a maximal ideal. Let $S = \{m\}$ in the lemma. Find $[\underline{C}] = x$ where $H(\underline{C}_m) = 0$. Hence $[\underline{C}_m] = 0$ but it is the image of x in $K_0(R_m)$. Thus $F^1K_0(R) \subset \widetilde{K_0(R)}$.

6) $F^1 \subset \widetilde{K}_0$.

We want to show that $x \in \widetilde{K}_0(R)$ implies the condition in lemma 10.3.

Let $x \in \widetilde{K_0(R)}$. Then in the map $K_0(R) \longrightarrow K_0(R_m) = Z$. x goes to 0. Then $x = [P] - [Q]$ goes to $[P_m] - [Q_m] = 0$. Since R_m is a local ring, P_m is isomorphic to Q_m. We need another lemma.

Lemma 10.4. Let R be any commutative ring and $S \subset$ m-spec(R) be a finite set. Let P and Q be finitely generated projective R modules such that P_m is isomorphic to Q_m for all $m \in S$. Then there exists $f \colon P \longrightarrow Q$ such that $f_m \colon P_m \longrightarrow Q_m$ is an iso for each $m \in S$.

First we finish the theorem assuming the lemma. $\underline{C}$ is $0 \longrightarrow \ldots \longrightarrow 0 \longrightarrow P \xrightarrow{f} Q \longrightarrow \ldots \longrightarrow 0$ where P is in an even position and Q is in an odd one. Then $[C] = [P] - [Q]$ and $[C_m] = 0$ for all $m \in S$. Therefore we have the condition of lemma 10.3. and $x \in F^1K_0(R)$.

Now to prove the lemma.

Proof. Let $g: P_m \longrightarrow Q_m$ be an iso.

$Hom_R(P, Q)_m$ is isomorphic to $Hom_{R_m}(P_m, Q_m)$ since P is finitely generated. So there is f/s which goes to g where $f \in Hom(P, Q)$ and $s \in R - m$. f goes to sg and sg is also an iso of P_m with Q_m. Hence there is $f: P \longrightarrow Q$ such that $f_m: P_m \longrightarrow Q_m$ is an iso. Let $s = \{m_1, \ldots, m_n\}$. Then there are $f_i: P \longrightarrow Q$ such that $f_{i_{m_i}} : P_{m_i} \longrightarrow Q_{m_i}$ are isos. By the Chinese Remainder Theorem there exist $a_i \in R$ such that

$$a_i = \begin{cases} 1 & \text{mod } m_i \\ 0 & \text{mod } m_j \quad i \neq j \end{cases} .$$

Let $f = \sum a_i f_i$. Then $f: P \longrightarrow Q$ and $f_{m_i}: P_{m_i} \longrightarrow Q_{n_i}$ is an iso mod $m_{i_{m_i}}$ and, hence is an iso by Nakayama's lemma.

Done with lemma 10.4 and theorem 10.2.

Corollary 10.5. Let R be a commutative noetherian ring with dim m-spec(R) = d. Then $\widetilde{K}_0(R)^{d+1} = 0$.

Corollary 10.6. Let R be any commutative ring. Then every element of $\widetilde{K}_0(R)$ is nilpotent.

Proof. Let $x \in \widetilde{K}_0(R)$. Then $x = [P] - [Q]$. We want to find a ring R' which is noetherian and has dim m-spec(R) finite and a map $R' \longrightarrow R$ such that x is in the image of $K_0(R') \longrightarrow K_0(R)$. Since P and Q are finitely generated projectives there exist finitely generated free modules F and F' and idempotent matrices $(a_{ij}) \in Hom(F, F)$ and $(b_{kl}) \in Hom(F', F')$ such that

$$\begin{array}{l} F \xrightarrow{(a_{ij})} F \longrightarrow P \longrightarrow 0 \quad \text{and} \\ F' \xrightarrow{(b_{kl})} F' \longrightarrow Q \longrightarrow 0 \quad \text{are} \end{array}$$

free resolutions of P and Q. Let x_{ij} and y_{kl} be indeterminates. Then

$$R' = Z[x_{ij}, y_{kl}]/(\text{relations coming from } (a_{ij})^2 = (a_{ij}) \text{ and } (b_{kl})^2 = (b_{kl}))$$

where x_{ij} goes to a_{ij} and y_{kl} goes to b_{kl}. Let P' be defined by the idempotent matrix (x_{ij}) and Q' by (y_{kl}). Tensoring with R sends P' to P and Q' to Q.

R' is generated by N elements over Z, therefore Krull dim $R' \le N + 1$. We have the commutative diagram

$$\begin{array}{ccccccccc} 0 & \longrightarrow & \widetilde{K_0(R')} & \longrightarrow & K_0(R') & \underset{r}{\overset{n}{\rightleftarrows}} & H_0(R') & \longrightarrow & 0 \\ & & \downarrow & & \downarrow & & \downarrow & & \\ 0 & \longrightarrow & \widetilde{K_0(R)} & \longrightarrow & K_0(R) & \longrightarrow & H(R) & \longrightarrow & 0 \end{array}.$$

We have $x \in \widetilde{K_0(R)} \subset K_0(R)$ and $y \in K_0(R')$ which goes to x. Let $z = y - nr(y)$. Then $r(z) = 0$ so $z \in \widetilde{K_0(R')}$ and z goes to $x - nr(x) = x$ since $x \in \widetilde{K_0(R)}$. Hence x is in the image of $\widetilde{K_0(R')} \longrightarrow \widetilde{K_0(R)}$. But $z^{d+1} = 0$ and $z^{d+1} \longrightarrow x^{d+1}$ where d is the dim m-spec(R'). Hence $x^{d+1} = 0$. Done.

<u>Remark</u>. We have shown that if R is any commutative ring and P and Q are finitely generated projective R modules then there exists $f: R' \longrightarrow R$ where R is a finitely generated z algebra and finitely generated projective R' modules P' and Q' such that $P' \otimes_{R'} R = P$ and $Q' \otimes_{R'} R = Q$. Hence for any $x \in K_0(R)$ we can find R' a finitely generated Z algebra such that $x \in$ image of $K_0(R') \longrightarrow K_0(R)$.

<u>Corollary 10.7</u>. $\widetilde{K_0}(R)$ = nil radical of $K_0(R)$ = Jacobson radical of $K_0(R)$ where R is any commutative ring.

<u>Proof</u>. $0 \to \widetilde{K_0} \to K_0 \to H \to 0$ is exact. H has no nil-potent elements. Hence $\widetilde{K_0}$ is the set of nil-potent elements.

Let $R \xrightarrow{f} R'$ be a homomorphism of rings which is onto. Then if J and J' are the Jacobson radicals $f(J) \subset J'$. For if M' is a maximal ideal of R' then $f^{-1}(M') = M$ is a maximal ideal of R. Therefore $J' = \bigcap_{M' \max} M'$. $f^{-1}(J') = \bigcap f^{-1}(M') \supset J$.

We need that H has 0 Jacobson radical.

Then $\widetilde{K_0} \supset J$ and $J \supset \widetilde{K_0}$. Let $x \in \operatorname{spec}(R)$. Then $H_0(R) \to Z$ by $f \in H(R)$ goes to $f(x)$ is onto. J = Jacobson radical of H goes to 0 = Jacobson radical of Z. Hence $f \in J$ implies $f(x) = 0$ for all $x \in \operatorname{spec}(R)$. That is $f = 0$. Hence $J = 0$. Done.

<u>Remark</u>. The theorem on nilpotence of $\widetilde{K_0}$ is motivated from topology. Let X be a finite connected complex. Then $K^0(X)$ is a ring and $\widetilde{K^0}(X)$ is nilpotent of degree equal to the dimension of X.

<u>Remark</u>. It is easy to extend the definition of F^i to $K_0(A)$ where A is a finite R-algebra. Proposition 10.2 still holds but 3 becomes $F^iK_0(A) \cdot F^jK_0(B) \subset F^{i+j}K_0(A \otimes_R B)$ so we can't conclude anything about nilpotence directly from this.

Chapter 11. Serre's Theorem.

In this chapter R is a commutative ring. A is an R algebra such that

1) m-spec(R) is a noetherian space of dim $d < \infty$, and

2) A is finitely generated as a R module (i.e., A is a finite R algebra).

If M is an A module can we write $M = A \oplus N$? In topological K-theory we have the result that if B is an n bundle over X, a finite complex of dim d and $n > d$, then $B = F \oplus N$ where F is a trivial $n - d$ bundle.

Serre's original theorem was: If P is a finitely generated projective R module and $r_m(P) > d$ for all $m \in$ m-spec(R), then $P = R \oplus Q$. We present a generalization due to Bass. Recall that a ring A with Jacobson radical J is semilocal if A/J has D.C.C.

Lemma 11.1. If A is a R algebra which is finitely generated as a R module, then for all $\underline{p} \in \text{spec}(R)$, $A_{\underline{p}}$ is semilocal and $\underline{p}A_{\underline{p}}$ is contained in the Jacobson radical of $A_{\underline{p}}$.

Proof. $A_{\underline{p}}$ is clearly finitely generated as a $R_{\underline{p}}$ module. Hence $A_{\underline{p}}/\underline{p}A_{\underline{p}}$ is a finitely generated $R_{\underline{p}}/\underline{p}R_{\underline{p}}$ module. But $R_{\underline{p}}/\underline{p}R_{\underline{p}}$ is a field, and hence we will be done if we show the last part of the lemma.

Let $x \in A_{\underline{p}}$ such that x is a unit mod $\underline{p}A_{\underline{p}}$. It is enough to show that x is a unit. We have $A_{\underline{p}} \xrightarrow{\hat{x}} A_{\underline{p}} \longrightarrow Q \longrightarrow 0$ where $\hat{x}(a) = ax$. Reduce mod $\underline{p}_{\underline{p}}$ yields

$$A_{\underline{p}}/\underline{p}A_{\underline{p}} \longrightarrow A_{\underline{p}}/\underline{p}A_{\underline{p}} \longrightarrow Q/\underline{p}Q \longrightarrow 0 .$$

But the left hand map is an iso. Therefore $Q/\underline{p}Q = 0$. But Ap finitely generated implies Q is finitely generated. Nakayama's lemma implies $Q = 0$. Hence x has a right inverse. Similarly it has a left inverse. Hence x is a unit and $\underline{p}A_{\underline{p}}$ is contained in the Jacobson radical of $A_{\underline{p}}$.

Done.

Definition. Let P be an A module and m a maximal ideal of R. Then $\text{f-rank}_m P \geq n$ means that P_m has a free direct summand with n free generators. That is $P_m = A_m^n \oplus Q$.

Theorem 11.2. Let M be an A module which is a direct summand of a direct sum of finitely presented A modules. If $\text{f-rank}_m M > d$ for all $m \in$ m-spec(R), then $M = A \oplus N$ for some N.

The proof of this theorem will take up most of the chapter. First we investigate when a mono splits.

Let S be a ring and M an S module. Then $M^* = \text{Hom}_S(M, S)$ is right S module. The S action is given by $s \in S$ and $f \in M^*$ then $fs(m) = f(m)s$. We have the natural map $M \longrightarrow M^{**}$ given by m goes to $\hat{m}$ where $\hat{m}(f) = f(m)$. It is clear that under this map $A \longrightarrow A^{**}$ is an iso. $\text{Hom}_S(\ , S)$ is an additive functor. Hence $Q \longrightarrow Q^{**}$ is an iso if Q is a finitely generated projective S module.

Definition. Let M be an S module. Then M is reflexive if the natural map $M \longrightarrow M^{**}$ is an iso.

It is clear that M and N are reflexive if and only if $M \oplus N$ is.

Lemma 11.3. Let S be a ring and Q a finitely generated projective S module. Let M be any S module and $f: Q \longrightarrow M$ a map of S modules. Then f is a split mono if and only if $f^*: M^* \longrightarrow Q^*$ is an epi.

Remark. Since Q^* is projective f^* will be a split epi if it is an epi. Lemma 11.3 would be obvious if every module were reflexive.

<u>Proof</u>. $\Longrightarrow$ This is trivial. Let $g: M \longrightarrow Q$ such that $gf = 1_Q$. Dualize and obtain $Q^* \underset{g^*}{\overset{f^*}{\rightleftarrows}} M^*$. Then $f^*g^* = 1_{Q^*}$ and f^* is an epi.

$\Longleftarrow$ Let $f^*: M^* \longrightarrow Q^*$ be an epi. Q^* is a (finitely generated) projective S module. Therefore, there exists $h: Q^* \longrightarrow M^*$ such that $f^*h = 1_{Q^*}$. Dualizing gives

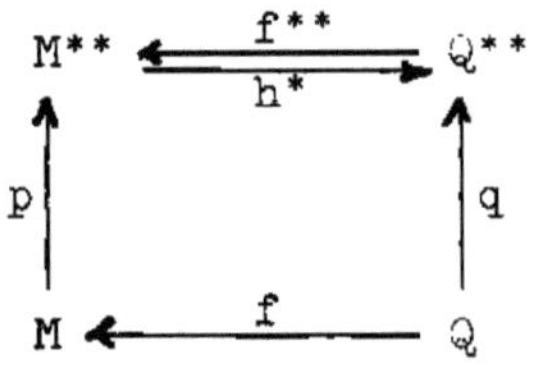

where q and p are the natural maps. Since Q is a finitely generated projective q is an iso. $h^*f^{**} = 1_{Q^{**}}$ since $f^*h = 1_Q$. Let $g = q^{-1}h^*p$. Then $gf = 1_Q$ if and only if $qgf = q$ since q is an iso. But $qgf = h^*pf = h^*f^{**}q = 1_{Q^{**}}q = q$. Hence f is a split mono. Done.

<u>Lemma 11.4</u>. Let S be a ring and Q a finitely generated projective S module. Let f and $g: Q \longrightarrow M$ where M is any S module. Let $\mathrm{im}(f - g) \subset JM$ where J is the Jacobson radical of S. Then f is a split mono if and only if g is.

<u>Proof</u>. Let f^* and g^* be the dual maps. Suppose that $\mathrm{im}(f^*-g^*) \subset Q^*J$. Then f^* will be an epi if and only if g^* is: Q^* is finitely generated. Hence by Nakayama's lemma f^* is an epi if and only if $f^*: M^*/M^*J \longrightarrow Q^*/Q^*J$ is. But if $\mathrm{im}(f^*-g^*) \subset Q^*J$, then f^* and g^* have the same image in Q^*/Q^*J. Hence, one is an epi if and only if the other is.

Let $h = f - g$ then $h^* = f^* - g^*$. We need to show that $h: Q \longrightarrow M$ such that $h(Q) \subset JM$ implies $h^*(M^*) \subset Q^*J$. Let $j: M \longrightarrow S$ be an element of M^*. $h^*(j) = jh$. If $h(Q) \subset JM$, then $jh(Q) = j(h(Q)) \subset j(JM) \subset J(j(M)) \subset JS$.

Thus we are reduced to proving that if Q is a finitely generated projective and a: $Q \longrightarrow S$ such that $a(Q) \subset J$, then $a \in Q^*J$. That is the function

$$\mathrm{Hom}_S(Q, J) \longrightarrow \mathrm{Hom}_S(Q, S) = Q^*$$

has image equal to Q^*J. This is clear if $Q = S$ since $\mathrm{Hom}_S(S, \_\_)$ is the identity functor. This is additive in Q. Therefore, since it is true for S it is true for any finitely generated free. But true for any finitely generated free implies it is true for $Q \oplus P$ where $Q \oplus P$ is finitely generated free. Additivity again implies it is true for Q. Done.

Lemma 11.5. Let A be an R algebra as above. Let Q be a finitely generated projective A module. Let M be a direct summand of a direct sum of finitely presented A modules. Let f: $Q \longrightarrow M$ be a map of A modules. Then

1) $\{\underline{p} \in \mathrm{spec}(R) \mid f_{\underline{p}}: Q_{\underline{p}} \longrightarrow M_{\underline{p}}$ is a split mono$\}$ is an open set of spec(R), and
2) If $f_m: Q_m \longrightarrow M_m$ is a split mono for all maximal ideals, m, of R, then f is a split mono.

Proof. There exists an A module N such that $M \oplus N = \coprod_{j \in J} M_j$ where M_j are all finitely presented. Then f is a split mono if and only if if is where i: $M \longrightarrow M \oplus N$ is the cannonical injection. Therefore we can assume that M is a direct sum of finitely presented A modules.

Since Q is finitely generated so is the image of f. Hence there is a finite set $S \subset J$ such that $f(Q) \subset \coprod_{j \in S} M_j$. Thus we can assume that M is a finite direct sum of finitely presented modules. That is, M itself is finitely presented.

We claim that $(M_{\underline{p}})^* = \mathrm{Hom}_{A_{\underline{p}}}(M_{\underline{p}}, A_{\underline{p}}) = \mathrm{hom}_A(M, A)_{\underline{p}} = (M^*)_{\underline{p}}$ for all $\underline{p} \in \mathrm{spec}(R)$. This is clear for M = A and, hence, for M any finitely generated free A module. But dual is a left exact functor and localizing is exact. Since M is finitely presented, there is an exact sequence $F \to F' \to M \to 0$ with F and F' finitely generated free A modules. We have the following commutative diagram.

$$\begin{array}{ccccccc} 0 & \to & (M^*)_{\underline{p}} & \to & (F'^*)_{\underline{p}} & \to & (F^*)_{\underline{p}} \\ & & \downarrow & & \downarrow & & \downarrow \\ 0 & \to & (M_{\underline{p}})^* & \to & (F'_{\underline{p}})^* & \to & (F_{\underline{p}})^* \end{array}$$

where the two right hand maps are isos. Hence, by the 5-lemma, the left hand map is an iso.

Therefore $f_{\underline{p}}\colon Q_{\underline{p}} \to M_{\underline{p}}$ is a split mono if and only if $(f^*)_{\underline{p}}\colon (M^*)_{\underline{p}} \to (Q^*)_{\underline{p}}$ is an epi. The sequence $(M^*)_{\underline{p}} \to (Q^*)_{\underline{p}} \to (\mathrm{coker}\, f^*)_{\underline{p}} \to 0$ is exact and $(f^*)_{\underline{p}}$ is an epi if and only if $(\mathrm{coker}\, f^*)_{\underline{p}} = 0$. But coker f^* is a finitely generated A module since Q is. Therefore, 1) is proved since supp(coker f^*) is closed.

If $f^*_{\underline{m}}$ is an epi for all m, then $(\mathrm{coker}\, f^*)_{\underline{m}}$ is 0 for all $m \in$ m-spec(R). Therefore coker $f^* = 0$ and f^* is an epi. Hence f is a split mono and 2) is proved. Done.

<u>Lemma 11.6</u>. (Chinese Remainder Theorem). Let M be any A module and let $m_1, \ldots, m_n$ be a finite number of maximal ideals of R with

$m_i \neq m_j$ if $i \neq j$. Let $a_1, \ldots, a_n$ be any n elements of M. Then there is an $a \in M$ such that $a \equiv a_i \bmod m_i M$.

Proof. We can find $x_i \in R$ such that

$$x_i \equiv 1 \bmod m_i \text{ and } x_i \equiv 0 \bmod m_j \quad i \neq j$$

by the usual Chinese Remainder Theorem. Then $a = \sum_{i=1}^{n} x_i a_i$ works for a.

Done.

Now we can prove case $d = 0$ of theorem 11.2. That is if A is a finitely generated R module which is an R algebra and m-spec(R) is a noetherian space of dim 0 and M is a direct summand of a direct sum of finitely presented A-modules with $\text{f-rank}_m M > 0$ for all $m \in$ m-spec(R), then $M = A \oplus N$.

Let X = m-spec(R). Then X is noetherian, T_1, and of dim 0. Therefore the irreducible components are points. Therefore $x = \{m_1, \ldots, m_n\}$ with the discrete topology. $\text{f-rank}_{m_i} M > 0$ hence $M_{m_i} = A_{m_i} \oplus N_i$. Let a_i/s_i be a generator of A_{m_i} where $a_i \in M$ and $s_i \in R - m_i$. Then $a_i/1$ also generates Am_i. By lemma 11.6 there is an $a \in M$ such that $a \equiv a_i \bmod m_i M$.

We claim a generates a free summand of M. That is we want that $f: A \longrightarrow M$ given by $f(r) = ra$ is a split mono. It is enough, by lemma 11.5.2 to show that $f_{m_i}: A_{m_i} \longrightarrow M_{m_i}$ is a split mono for all i. We are given that $g: A_{m_i} \longrightarrow M_{m_i}$ by $g(r) = ra_i$ is a split mono. But $\text{im}(f - g) \subset m_{i_{m_i}} M_{m_i} \subset$ (Jacobson radical of A_{m_i})M_{m_i} by lemma 11.1. Therefore, by lemma 11.4 f is a split mono for all i. Therefore, f is a split mono and case 0 of theorem 11.2 is done.

<u>Proposition 11.7</u>. Let A be a semi local ring. Let M and N be A modules such that $A \oplus M$ is isomorphic to $A \oplus N$. Then M is isomorphic to N.

<u>Proof</u>. First we need a lemma due to Bass.

<u>Lemma 11.8</u>. Let A be a semi-local ring and let $u \in A$ and $\underline{a}$ a left ideal of A such that $A = Au + \underline{a}$. Then there exists an $a \in \underline{a}$ such that u + a is a unit.

<u>Proof</u>. Let J = Jacobson radical of A and let $\bar{A} = A/J$. Then $\bar{A} = \overline{Au} + \bar{\underline{a}}$. If $\bar{u} + \bar{a}$ is a unit in $\bar{A}$, then given any u + a that has image $\bar{u} + \bar{a}$ we know that u + a is a unit. Hence we can assume that J = 0 and that A is semi simple. By the elementary properties of semi simple rings, there is a left ideal $\underline{b} \subset \underline{a}$ such that $A = Au \oplus \underline{b}$. The sequence $0 \longrightarrow \underline{c} \longrightarrow A \longrightarrow Au \longrightarrow 0$ where $A \longrightarrow Au$ is given by $x \longrightarrow xu$ splits. Let $f: A \longrightarrow \underline{c}$ be a splitting. Also $\underline{c}$ is isomorphic to $\underline{b}$. Let $g: \underline{c} \longrightarrow \underline{b}$ be an iso. Then $b = g(f(1)) \in \underline{b}$ and u + b is a right unit since the composition of the isos $A \xrightarrow{(u,\, f)} Au \oplus \underline{c} \xrightarrow{1 \oplus g} Au \oplus \underline{b} \xrightarrow{=} A$ sends 1 to u + b. But by the elementary properties of semi simple rings u + b is then a unit.

Done with lemma 11.8.

We have an exact sequence

$0 \longrightarrow N \longrightarrow A \oplus M \longrightarrow A \longrightarrow 0$ coming from the direct sum decomposition of $A \oplus N$ and the iso of $A \oplus N$ and $A \oplus M$. This gives maps $f: M \longrightarrow A$ and $\hat{a}: A \longrightarrow A$ given by $\hat{a}(1) = 1a$ for every $1 \in A$. (Every map $A \longrightarrow A$ has this form.) Since the map $A \oplus M \longrightarrow A$ is onto, $A = Aa + f(M)$. Therefore, by lemma 12.2 there is a unit u of A such that $u = a + f(m)$ for some $m \in M$. We define the map

$h: A \oplus M \longrightarrow A \oplus M$ by the composition of the two maps

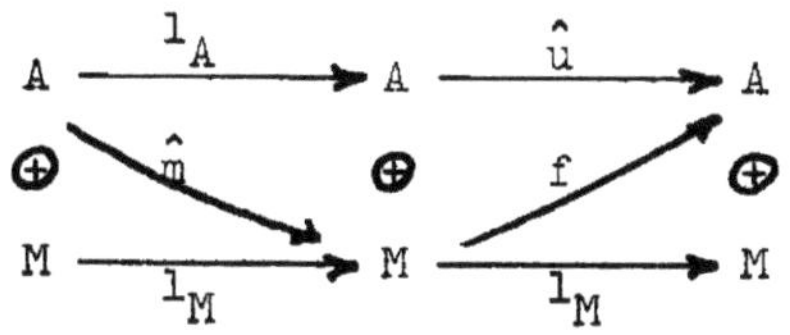

where $\hat{m}(r) = rm$ and $\hat{u}(r) = ru$. Then h is an iso since it is the composition of two isos and $h(r, x) = (ru + f(x - rm), x - rm)$. But $ra + f(x) = r(u - f(m)) + f(x)$ since $a = u - f(m)$. Therefore the diagram

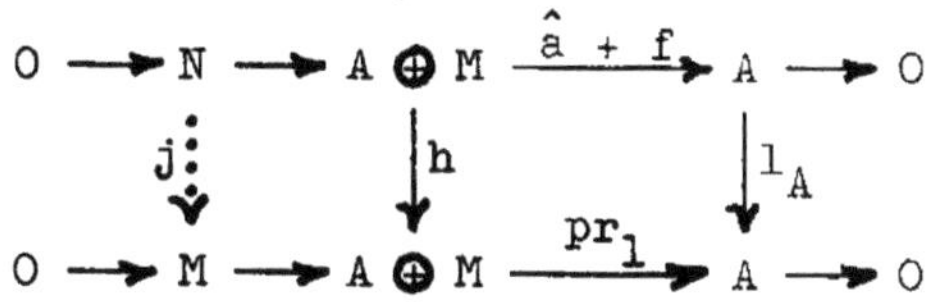

$$\begin{array}{ccccccccc} 0 & \longrightarrow & N & \longrightarrow & A \oplus M & \xrightarrow{\hat{a} + f} & A & \longrightarrow & 0 \\ & & {\scriptstyle j} \downarrow & & {\scriptstyle h} \downarrow & & {\scriptstyle 1_A} \downarrow & & \\ 0 & \longrightarrow & M & \longrightarrow & A \oplus M & \xrightarrow{pr_1} & A & \longrightarrow & 0 \end{array}$$ commutes.

Therefore there is a map $j: N \longrightarrow M$ making the diagram commute and j is an iso since h and 1_A are by the 5 lemma.

Done with Proposition 11.7.

If we think of $f\text{-rank}_m M$ to be a function of m and say that $f\text{-rank}_m M = n$ if $f\text{-rank}_m M \geqslant n$ but $f\text{-rank}_m M \not\geqslant n + 1$ and $f\text{-rank}_m M = \infty$ if $f\text{-rank}_m M \geqslant n$ for all n, then we can state the corollary:

<u>Corollary 11.9</u>. Let M be any A module. Then $f\text{-rank } A \oplus M = 1 + f\text{-rank } M$.

<u>Proof</u>. For every $m \in \text{m-spec}(R)$ A_m is a semi local ring by lemma 11.1. It is clear that $f\text{-rank } A \oplus M \geqslant 1 + f\text{-rank } M$ for all $m \in \text{m-spec}(R)$. Suppose $(A \oplus M)_m$ has a free summand on r generators then $(A \oplus M)_m \simeq A_m^r \oplus N \simeq A_m \oplus A_m^{r-1}$. Therefore, by proposition 12.1

$M_m \cong A_m^{r-1} \oplus N$. Hence f-rank $A \oplus M \leq 1 +$ f-rank M. Done.

Definition. Let M be an A module, $s_1, \ldots, s_n \in M$, and $m \in$ m-spec(R). Then $s_1, \ldots, s_n$ are good at m if $s_1, \ldots, s_n$ are a basis for a free summand of M_m.

By corollary 11.9 and the case d = 0 of theorem 11.2 we see that if $s_1, \ldots, s_n$ are good at m and f-rank$_m$M $> n$, then we can find a $w \in M$ so that $s_1, \ldots, s_n$, w are good at m.

Let $s_1, \ldots, s_n \in M$, by lemma 11.5 the bad set, {m where $s_1, \ldots, s_n$ are not good}, is closed and if the bad set is empty then $s_1, \ldots, s_n$ are a basis for a free summand of M.

We want to find an $s \in M$ such that the s is good for all $m \in$ m-spec(R). If $s_1, \ldots, s_n \in M$ are good at m and $t_1, \ldots, t_n \in M$ are such that $t_i \equiv s_i$ mod m M then $t_1, \ldots, t_n$ are good at m by lemmas 11.1 and 11.4. f-rank$_m$M $\geq n$ if and only if there exist $s_1, \ldots, s_n \in M$ which are good at m.

The case of d = 0 has already been done. Next we prove a lemma which gives us the appropriate statement to use induction on.

Lemma 11.10. Let M be an A module which is a direct summand of a direct sum of finitely presented A modules and let m-spec(R) be a noetherian space of dim $d < \infty$. Let f-rank$_m$M $\geq r$ for all $m \in$ m-spec(R). Suppose we are given:

1) a closed set F of m-spec(R),
2) $s_1, \ldots, s_h \in M$ which are good outside of F,
3) $x_1, \ldots, x_n \in F$ and $v_1, \ldots, v_n \in M$, and
4) $k \leq r - h$.

Then there exists $s \in M$ and a closed set F' such that

a) $s_1, \ldots, s_h$, s are good outside of $F \cup F'$;

b) $s_i \equiv v_i \bmod x_i M$ for all i; and

c) codim $F' \geq k$.

Remark. Lemma 11.10 implies theorem 11.2. Take h = 0, n = 0, and $F = \emptyset$. Then $r > d$ by hypothesis and hence $k = d + 1 \leq r$. Therefore,

a) There exists $s \in M$ good outside of F';

b) is vacuous; and

c) codim $F' \geq d + 1$.

But dim m-spec(R) = d. Hence $F' = \emptyset$. Hence s is good everywhere and the submodule (s) splits off as a free summand of M. That is, $M = A \oplus N$.

Proof of lemma 11.10. First we do the case k = 0. Let F' = m-spec(R).

a) is vacuous;

b) follows by the Chinese Remainder Theorem; and

c) is vacuous.

We proceed by induction on k. Assume the lemma is true for k - 1 where $k \geq 1$. We want to prove it for k. By assumption there exists $s \in M$ such that $s_1, \ldots, s_h$, s are good outside of $F \cup G$; $s \equiv v_i \bmod x_i M$; and codim $G \geq k - 1$. The problem is that G is too big. Let $G_1, \ldots, G_m$ be the irreducible components of G which are not contained in F. Pick $y_i \in G_i$ such that $y_i \notin F \cup \bigcup_{j \neq i} G_j$. We use induction on the data

1) $F \cup G$,

2) $s_1, \ldots, s_h$, s,

3) $x_1, \ldots, x_n, y_1, \ldots, y_m \in F \cup G$ and
$0_1, \ldots, 0, w_1, \ldots, w_m \in M$ (to be specified below), and
4) $k - 1 \leqslant r - (h + 1)$ is satisfied since $k \leqslant r - h$.

Pick w_j such that $s_1, \ldots, s_h, s + w_j$ are good at y_j. As we observed above, this is possible since $\text{f-rank}_{y_j} M \geqslant r \geqslant h + 1$. There is a closed set H and a $u \in M$ such that codim $H \geqslant k - 1$, $s_1, \ldots, s_h, u$ are good outside of $F \cup G \cup H_j$, and $u \equiv 0 \bmod x_i M$ and $\equiv w_j \bmod y_j M$.

We claim there is an $a \in R$ such that $s + au$ satisfies the lemma. That is $s_1, \ldots, s_h, s + au$ are good outside of $_F \cup F'$, codim $F' \geqslant k$, and $s + au \equiv v_i \bmod x_i M$.

Let $\{H_b\}$ be the set of components of H which are not contained in $F \cup G$. Pick $z_b \in H_b$ such that $z_b \notin F \cup G \cup \bigcup_{b \neq b'} H_{b'}$. Pick $a \in R$ such that

$$a \equiv \begin{cases} 0 \bmod z_b & \text{for all } b \\ 1 \bmod y_j & \text{for all } j. \end{cases}$$

Then $s_1, \ldots, s_h, s + au$ are good outside of $F \cup G \cup H$ since $s_1, \ldots, s_h, s + w_j$ are good at y_j and good at z_b since $s_1, \ldots, s_h, s$ are good outside of $F \cup G$.

Let B be the bad set of $s_1, \ldots, s_h, s + au$. Then $B \subset F \cup G \cup H - \{y_j, z_b\}$. Let B_i be the irreducible components of B and let F' be the union of those B_i not contained in F. Then F' is closed and $B = F \cup F'$. We claim codim $F' \geqslant k$. It will suffice to show codim $B_i \geqslant k$ for $B_i \not\subset F$. Since $B \subset F \cup G \cup H$, we see that $B_i \subset G_j$ or $B_i \subset H_b$ for some $G_j \not\subset F$ or $H_b \not\subset F \cup G$. (If $H_b \subset F \cup G$, then $B_i \subset G$ so $B_i \subset G_j$). The y_j's and z_b's are not in B_i. Therefore, $B_i < H_b$ or $B_i < G_j$. Therefore, codim $F' \geqslant k - 1 + 1 = k$ as desired.

Done with lemma 11.10 and theorem 11.2.

Corollary 11.11. Let k be a field and $x_1, \ldots, x_d$ be indeterminates. Then m-spec$(k[x_1, \ldots, x_d])$ has dim d. If P is a finitely generated projective $k[x_1, \ldots, x_d]$ module and if $\text{f-rank}_m P \geq d + 1$ for all $m \in$ m-spec $(k[x_1, \ldots, x_d])$, then $P = R \oplus P'$.

Remark. Since $k[x_1, \ldots, x_n]$ is a domain $\text{f-rank}_m P$ is a constant function from m-spec to z.

Examples. 1) Let R be a commutative ring and A an R algebra which is finitely generated as an R module. Let m-spec(R) be a noetherian space of dim $d < \infty$. Then

$$\widetilde{K_0(A)} = \bigcap_{\underline{p} \in \text{spec}(R)} \ker[K_0(A) \longrightarrow K_0(A_{\underline{p}})]$$

$$= \bigcap_{m \in \text{m-spec}(R)} \ker[K_0(A) \longrightarrow K_0(A_m)].$$

Pick $x \in K_0(A)$. Then $x = [P] - [Q]$ where P and Q are finitely generated projective A modules. There exists a finitely generated projective A module, Q', such that $Q \oplus Q'$ is free. $x = [P \oplus Q'] - [Q \oplus Q']$. Hence any element in $K_0(A)$ has the form $[P] - [F]$ where P is a finitely generated projective and F is a finitely generated free. Suppose $x \in \widetilde{K_0(A)}$. Then $x = [P] - [F]$ and, in $K_0(A_m)$, $[P_m] - [F_m] = 0$. Therefore, there exists a finitely generated free F' such that $P_m \oplus F'_m \cong F_m \oplus F'_m$. A_m is semi-local. Therefore by proposition 11.7 $P_m \cong F_m$. If $F = A$, then P_m is free on r generators. If $r > d$, then Serre's theorem implies that $P = A \oplus P'$. Therefore $x = [P] - [A^r] = [P'] - [A^{r-1}]$. Thus we obtain:

Corollary 11.12. Let R be a commutative ring and A an R algebra which is finitely generated as an R module. Let $\dim$ m-spec(R) = $d < \infty$. Then every element of $\widetilde{K}_0(A)$ has the form $[P] - [A^r]$ where P is a finitely generated projective and $r \leq d$.

2) Let G be a finite group. Let $A = Z[G]$ be its integral group ring. Pick $x \in \widetilde{K}_0(Z[G])$. Then by corollary 11.9. $x = [P]$ or $x = [P] - [Z[G]]$. In the first case we get $P_m = 0$ for all $m \in$ m-spec(Z) and therefore $P = 0$. In the second case as an abelian group P is free and is generated by g elements where g is the order of G. Using the theorem of Jordan which states that there are only finitely many isomorphism classes of Z[G] modules which are free on a given number of generators as an abelian group, we obtain that $\widetilde{K}_0(Z[G])$ is finite. It can be shown that $K_0(Z[G]) \cong Z \oplus \widetilde{K}_0(Z[G])$.

3) Let R be a commutative noetherian ring of dim 1 (e.g., R Dedekind). If $x \in \widetilde{K}_0(R)$, then $x = [P] - [R]$ where P is a finitely generated rank 1 projective. We have the composition of functions

$$\mathrm{Pic}(R) \longrightarrow \widetilde{K}_0(R) \xrightarrow{\det} \mathrm{Pic}(R)$$

$$[P] \rightsquigarrow [P] - [R] \rightsquigarrow [P][R]^{-1} = [P].$$

Serre's theorem says $\mathrm{Pic}(R) \longrightarrow \widetilde{K}_0(R)$ is onto. $\mathrm{Pic}(R) \longrightarrow \widetilde{K}_0(R)$ is one-one since the composite is one-one. Therefore, det is an iso of Pic(R) and $\widetilde{K}_0(R)$.

If P and $Q \in \mathrm{Pic}(R)$, then $[P \otimes Q] \longrightarrow [P \otimes Q] - [R]$ and $[P] \longrightarrow [P] - [R]$ and $[Q] \longrightarrow [Q] - [R]$ in the maps $\mathrm{Pic}(R) \longrightarrow K_0(R)$. Therefore $[P \otimes Q] - [R] = [P] + [Q] - [R] - [R]$ and $R \oplus (P \otimes Q)$ is isomorphic to $P \oplus Q$. If P is a rank 1 projective then $P \cong \underline{a}$ where a is an ideal of R. Therefore we have proved that if R is a commutative noetherian ring of dim 1 and a and b are projective ideals, then $a \oplus b \cong R \oplus ab$.

Chapter 12. Cancellation Theorems.

In this chapter we investigate the following: Let R be a ring and A, B, and C be R modules such that $A \oplus B$ is isomorphic to $C \oplus B$. Is A isomorphic to C? It is easy to see this is false if we don't restrict to finitely generated modules even for R a field.

Example. Let A be a semi local ring, Q a finitely generated projective A module, and M and N any A modules. If $M \oplus Q$ is isomorphic to $N \oplus Q$, then M is isomorphic to N.

Proof. There exists an A module P such that $Q \oplus P$ is isomorphic to A^n where n is finite. Then $M \oplus A^n \cong M \oplus P \oplus Q \cong N \oplus P \oplus Q \cong N \oplus A^n$ and we apply proposition 11.7. to cancel the A's one at a time.

Done.

Definition. Let A be a ring and M an A module. Then $m \in M$ is unimodular if m is a base for a free direct summand of M.

Remarks. $m \in M$ is unimodular if and only if there is a map $f: M \longrightarrow A$ such that $f(m) = 1$. Let $h: M \longrightarrow M$ be an iso. Then $m \in M$ is unimodular if and only if $h(m)$ is.

Let $a \in A^r$ be unimodular and let $P = A^r/Aa$. Then $A^r \cong A \oplus P$. We can ask if $P \cong A^{r-1}$. That is is P free on r - 1 generators. Let A^r be free on $e_1, \ldots, e_r$ and P be free on $e_2', \ldots, e_r'$. Then there is an automorphism $h: A^r \longrightarrow A^r$ such that $h(e_1) = a$ and $h(e_i) = e_i'$ if $2 \leq i \leq r$. If there is an automorphism $h: A^r \longrightarrow A^r$ such that

$h(e_1) = a$ and $h(e_i) \in P$ for $2 \leqslant i \leqslant r$ where we think of P or a submodule of A^r via the iso $A^r \cong A \oplus P$, then P is clearly free on $h(e_i)$ for $2 \leqslant i \leqslant r$.

Proposition 12.1. Let P and a be as above. Then P is free on r - 1 generators if and only if there is an automorphism $h: A^r \longrightarrow A^r$ such that $h(e_1) = a$.

Proof. $\Longrightarrow$ $P \cong A^r/Aa \cong h(A^r)/h(e_1) \cong h(A^r/Ae_1) \cong A^{r-1}$.

$\Longleftarrow$ is clear from the above. Done.

Remark. Let $h: A^r \longrightarrow A^r$ be the above automorphism. Then $a = h(e_1) = \sum a_{1j}e_j$. Thus, we obtain if $a = \sum a_{1j}e_j$ and $P = A^r/Aa$ Then P is free if and only if $(a_{11}, \ldots, a_{1n})$ can be completed to a $n \times n$ invertible matrix. Since a is unimodular there is a map $f: A^r \longrightarrow A$ such that $f(a) = 1$. Therefore, $1 = f(a) = f(\sum a_{ij}e_j) = \sum a_{1j}f(e_j)$. Thus $a \in A^r$ is unimodular if and only if there exist $b_j \in A$ such that $\sum a_jb_j = 1$ where $a = \sum a_{1j}e_j$ in terms of a basis $\{e_j\}$. This is Serre's unimodular row problem. That is can every unimodular row, (a_{1j}) such that there exist b_j with $\sum a_{1j}b_j = 1$, be extended to an invertible matrix. Equivalently, given a unimodular row (a_j) is there an invertible matrix C such that $(a_1, \ldots, a_n)\, C = (1, 0, \ldots, 0)$. We can weaken this to finding an invertible matrix C such that $(a_1, \ldots, a_n)\, C = (a_1', \ldots, a_{n-1}', 0)$. If we could always remove one non zero term at a time we could end up with $(a_1, \ldots, a_n)\, C = (a, 0, \ldots, 0)$ where a is a unit. Then Ca^{-1} would send (a_j) to $(1, 0, \ldots, 0)$. The following result shows one step is enough.

<u>Proposition 12.2</u>. Let A be a ring and $a = (a_1, \ldots, a_n)$ be a unimodular row. Let $A^r = Aa \oplus P$. Then P is isomorphic to A^{r-1} if and only if there is an invertible $r \times r$ matrix C such that $(a_1, \ldots, a_r) C = (a_1', \ldots, a_{r-1}', 0)$.

<u>Proof</u>. $\Leftarrow$ is clear.

$\Rightarrow$ Let $C = (c_{ij})$. Then $a_j' = \sum_{i=1}^{r} a_i c_{ij}$ and $a_r' = 0$.

Let $\{e_i'\}$ be the image of the cannonical base under C. Then $a = \sum_{j=1}^{r} a_j' e_j' = \sum_{j=1}^{r} a_j' e_j'$ since $a_r' = 0$. Therefore, a is a unimodular row of A^{r-1} in terms of the new basis. Therefore, Aa is a direct summand of A^{r-1}. Say $A^{r-1} = Aa \oplus Q$. Then $A^r = (Aa \oplus Q) \oplus A =$ $= Aa \oplus (Q \oplus A) = Aa \oplus A^{r-1}$. Therefore $P = A^r/Aa \cong A^{r-1}$. Done.

<u>Definition</u>. Let e_i be a basis for A^r. Then C is an <u>elementary transformation</u> if C sends e_i to e_i for all i except one say j and C send e_j to $e_j + se_k$ where $s \in A$ and $k \neq j$. C is clearly invertible.

If $(a_1, \ldots, a_r)$ is any element of A^r then $(a_1, \ldots, a_r) C = (a_1, \ldots, a_j + sa_k, \ldots, a_r)$. We can ask if we can send a unimodular row $(a_1, \ldots, a_r)$ to $(a_1', \ldots, a_{r-1}', 0)$ by a sequence of elementary transformation. That is a stronger requirement than in proposition 12.2 but it has an easily formulatable answer.

<u>Theorem 12.3</u>. (Bass). Let R be a commutative ring, A an R algebra which is finitely generated an R module. Suppose m-spec(R) is a noetherian space of dim $d < \infty$. Then if $(a_1, \ldots, a_r) \in A^r$ is a unimodular row and $r \geq d + 2$, then there exist $b_1, \ldots, b_{r-1} \in A$ such that $(a_1 + a_r b_1, \ldots, a_{r-1} + a_r b_{r-1})$ is a unimodular row of A^{r-1}.

Remark. This is the same as what we want for if $(a_1, \ldots, a_r) \rightsquigarrow (a_1', \ldots, a_{r-1}', a_r)$ where $(a_1', \ldots, a_{r-1}')$ is unimodular then we can find $c_i \in A$ such that $\sum_{i=1}^{n-1} a_i'c_i = 1$ and $\sum a_i'c_ia_r = a_r$. Therefore we can use more elementary transformations to send $(a_1', \ldots, a_{r-1}', a_r)$ to $(a_1', \ldots, a_{r-1}', 0)$.

Corollary 12.4. Let R and A be as above. If $A \oplus P$ is isomorphic to A^r where $r \geq d + 2$, then P is isomorphic to A^{r-1}.

Proof of corollary from the theorem is clear from proposition 12.2.

Before we proved theorem 12.3, we want to generalize it. Let $f: A \longrightarrow A^{r-1}$ be given by $f(1) = (b_1, \ldots, b_{r-1})$. Then $f(a_r) = (a_rb_1, \ldots, a_rb_{r-1})$. Theorem 12.7 then states that if we are given $a = (a_1, \ldots, a_r) \in A^{r-1} \oplus A$ which is unimodular then there exists $f: A \longrightarrow A^{r-1}$ such that $a - f(a_r)$ is unimodular in A^{r-1} when $r \geq d + 2$. This leads to the generalization of theorem 12.3.

Theorem 12.5. Let R be a commutative ring, let A be an R algebra which is finitely generated as an R module, and let m-spec(R) be a noetherian space of dim $d < \infty$. If Q and P are projective A modules with $\text{f-rank}_m P > d$ for all $m \in$ m-spec(R), M is any A module, and $a = (a_Q, a_P, a_M) \in Q \oplus P \oplus M$ is unimodular then there exists an $f: Q \longrightarrow P$ such that $(a_P + f(a_Q), a_M)$ is unimodular in $P \oplus M$.

Corollary 12.6. Let A, R, P, Q, and M be as above. If Q is finitely generated and $Q \oplus P \oplus M$ is isomorphic to $Q \oplus N$ where N is any A module, then $P \oplus M$ is isomorphic to N.

Proof of corollary 12.6 from theorem 12.5. We can find Q' such that $Q \oplus Q' = A^n$ $n < \infty$. Then $A^n \oplus P \oplus M$ is isomorphic to $A^n \oplus N$.

Hence we only need to cancel A one at a time. Let $a = (a_A, a_P, a_M)$ be the image of (1, 0) under an iso $h: A \oplus N \longrightarrow A \oplus P \oplus M$. a is unimodular since (1, 0) is. By the theorem there is an $f: A \longrightarrow P$ such that $(a_P + f(a_A), a_M)$ is unimodular in $P \oplus M$. We define $g: A \oplus P \oplus M \longrightarrow A \oplus P \oplus M$ by the diagram:

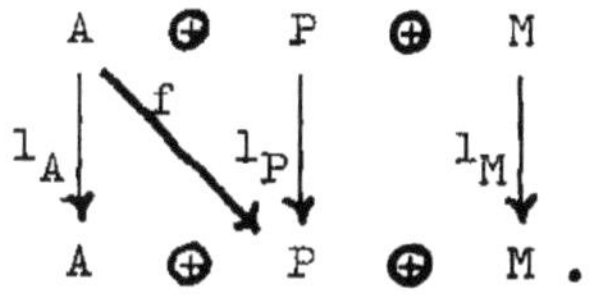

Then g is clearly an iso and $g((a_A, a_P, a_M)) = (a_A, a_P + f(a_A), a_M)$. Hence, using the iso, $gh: A \oplus N \longrightarrow A \oplus P \oplus M$ we can assume that $a = (a_A, a_P, a_M) = gh((1, 0))$ where (a_P, a_M) is unimodular. Therefore, there exists $j: P \oplus M \longrightarrow A$ such that $j((a_P, a_M)) = 1$. Define $k: A \longrightarrow A$ by $k(1) = a_A$. Now we define $l: A \oplus P \oplus M \longrightarrow A \oplus P \oplus M$ by the diagram

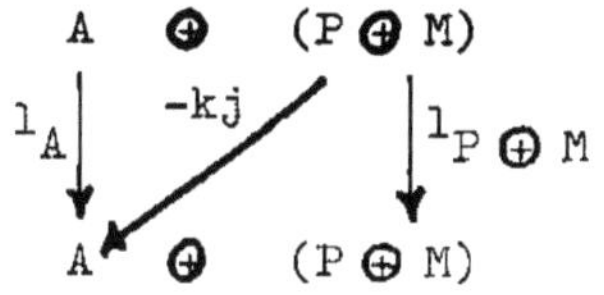

$l((a_A, a_P, a_M)) = (a_A - a_A, a_P, a_M)$ and l is an iso. Therefore we can assume that in the iso $i: A \oplus N \longrightarrow A \oplus P \oplus M$ that $i((1, 0)) =$ $= (0, a_P, a_M)$. That is we can assume $a_A = 0$. Let $M' = P \oplus M$. We want to show that M' is isomorphic to N. Since $a = (a_P, a_M)$ is unimodular in M', we have $M' = Aa \oplus M''$ for some M''. Let $X = A \oplus (Aa \oplus M'')$. Then X/Aa is isomorphic by $i - 1$ to $(A \oplus N)/A((1, 0)) \cong N$. But X/Aa is also isomorphic to $A \oplus M''$ which is isomorphic to $Aa \oplus M'' = M'$. Therefore $N \cong M'$.

Done with corollary 12.6 modulo theorem 12.5.

<u>Definition</u>. Let A be any ring, N an A module, and $n \in N$. Then $O_N(n) = \{f(n) | f: N \longrightarrow A\}$.

<u>Remarks</u>. $O_N(n)$ is a right ideal of A. n is unimodular if and only if $1 \in O_N(n)$, i.e., if and only if $O_N(n) = A$. If $N = N' \oplus N''$ and $n = (n', n'')$, then $O_N(n) = O_{N'}(n') + O_{N''}(n'')$.

We can restate theorem 12.5 in terms of O's. We have $a \in Q \oplus P \oplus M$ unimodular so $O_Q(a_Q) + O_P(a_P) + O_M(a_M) = A$ and we want an $f: Q \longrightarrow P$ such that $O_P(a_P + f(a_Q)) + Q_M(a_M) = A$. In this formulation $O_M(a_M)$ does not change. Hence we can reformulate theorem 12.5 as follows:

<u>Theorem 12.7</u>. Let R, A, d, P, and Q be as in theorem 12.5. Let $\underline{a}$ be a right ideal of A. If $a_P \in P$ and $a_Q \in Q$ such that $O_Q(a_Q) + O_P(a_P) + \underline{a} = A$, then there exists $f: Q \longrightarrow P$ such that $O_P(a_P + f(a_Q)) + \underline{a} = A$.

<u>Remark</u>. This appears to be more general than theorem 12.6. However it is an easy consequence of that theorem. Suppose $1 = x + y + z$ with $x \in O_Q(a_Q)$, $y \in O_P(a_P)$, $z \in \underline{a}$. Then $O_a(a_Q) + O_P(a_P) + zA = A$. But zA has the form $O_M(a_M)$. In fact if $z \in A$ then $O_A(z) = zA$.

Let A be a ring and I a 2 sided ideal of A. Let $^-$ denote the image under the cannonical map $A \longrightarrow A/I$.

<u>Lemma 12.8</u>. If P is a projective A module and $a \in P$, then $O_{\overline{P}}(\overline{a}) = \overline{O_P(a)}$ where $\overline{P} = P/IP$.

<u>Proof</u>. Let $f: P \longrightarrow A$ with $f(a) = r$. That is $r \in O_P(a)$.

Then we have

$$\begin{array}{ccc} P & \xrightarrow{f} & A \\ \downarrow & & \downarrow \\ \bar{P} & \xrightarrow{\bar{f}} & \bar{A} \end{array}$$ is a commutative

diagram, and $\bar{f}(\bar{a}) = \bar{r}$. Therefore, $O_{\bar{P}}(\bar{a}) \supset \overline{O_P(a)}$.

Let $g: \bar{P} \longrightarrow \bar{A}$. Then we have the diagram of A modules

$$\begin{array}{ccccc} & & P & & \\ & \swarrow^{h} & \downarrow & & \\ & & \bar{P} & & \\ & & \downarrow g & & \\ A & \longrightarrow & \bar{A} & \longrightarrow & 0 . \end{array}$$

But P is projective. Therefore, there is an $h: P \longrightarrow A$ making the diagram commute. Therefore $h(a) = s$ where $\bar{s} = g(\bar{a})$. Hence, $O_{\bar{P}}(\bar{a}) = \overline{O_P(a)}$. Done with lemma 12.8.

Now we begin proving theorem 12.7. by constructing f a little at a time.

Fix Q, P, and $\underline{a}$. Let $g: Q \longrightarrow P$ then the theorem is true for (a_P, a_Q) if and only if it is true for $(a_P + g(a_Q), a_Q)$. For if $f: Q \longrightarrow P$ works for (a_P, a_Q) then $f - g$ works for $(a_P + g(a_Q), a_Q)$. The converse follows by symmetry. Hence the conclusion is unchanged. $O_{P \oplus Q}((a_P, a_Q)) = O_P(a_P) + O_Q(a_Q)$ (and $O_{P \oplus Q}((a_P + g(a_Q), a_Q) = O_Q(a_Q) + O_P(a_P + g(a_Q))$). If these two are the same the hypothesis will also be unchanged. Let $h: P \oplus Q \longrightarrow P \oplus Q$ be defined by

$$\begin{array}{ccc} P & \oplus & Q \\ 1_P \downarrow & \swarrow g & \downarrow 1_Q \\ P & \oplus & Q \end{array} .$$

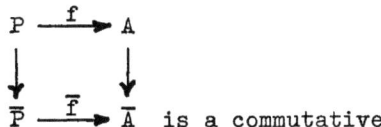

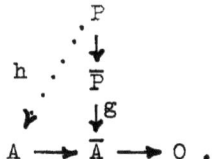

Then h is an iso, $h(a_P, a_Q) = (a_P + g(a_Q), a_Q)$, and $O(a) = O(h(a))$. Therefore, the hypothesis is unchanged.

Let $X = \text{m-spec}(R)$. Then $X = X_1 \cup \ldots \cup X_n$ where the X_i are the maximal irreducible components. Pick $m_i \in X_i$. Let $\underline{c} = m_1 \cap \ldots \cap m_n$. Then $R/\underline{c} = \prod R/m_i$ by Chinese Remainder Theorem. $A/\underline{c}A = \prod A/m_i A = \prod R/m_i \otimes_R A$. Hence $A/\underline{c}A$ has D.C.C. Therefore, $A/\underline{c}A$ is semi-local.

Serre's theorem applies to P. Say $P = A \oplus P'$ and $a_P = (b, a')$. Then $O_P(a_P) = O_A(b) + O_{P'}(a')$. But $O_A(b) = bA$. Therefore, $O(a_Q) + bA + Q(a') + \underline{a} = A$. Reducing mod $\underline{c}$ we get

$$\overline{O_Q(a_Q)} + \overline{bA} + \overline{O_{P'}(a')} + \overline{\underline{a}} = \overline{A} .$$

By lemma 11.8, there exist $\overline{q}' \in \overline{O_Q(a_Q)}$, $\overline{P}' \in \overline{O_{P'}(a')}$, and $\overline{a}'' \in \overline{\underline{a}}$ such that $\overline{u} = \overline{q'} + \overline{p'} + \overline{a''} + \overline{b}$ is a unit in $A/\underline{c}A$. We lift to $u = q' + p' + a'' + b$, choosing $p' \in O_{P'}(a')$, $q' \in O_Q(a_Q)$

1) We can assume that $p' = 0$. For $p' \in O_{P'}(a')$. Hence there exists $g: P' \longrightarrow A$ such that $g(a') = p'$. We construct $h: P \longrightarrow P$ by the diagram

$$\begin{array}{ccc} A & \oplus & P' \\ 1_A \downarrow & \swarrow g & \downarrow 1_{P'} \\ A & \oplus & P' \end{array} .$$

h is an iso. Then $a_P = (b, a')$ goes to $(b + p', a')$. Hence b in the above is replaced by $b + p'$, and $u = q' + (b + p') + a''$.

2) We can assume that $q' = 0$. In fact $q' \in O_Q(a_Q)$. Therefore, there exists $q: Q \longrightarrow A$ such that $g(a_Q) = q'$. We define

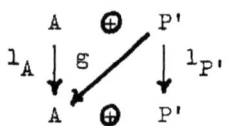

: $Q \longrightarrow P = A \oplus P'$ by $h = (g, 0)$. Then we replace a_P by $_P + h(a_Q) = a_P + q' = (b + q', a')$. Then b is replaced by $b + q'$. herefore $0 + b + q' + 0 + a'' = u$ where $b + q' \in O_A(b + q')$. Hence oth p' and q' can be assumed 0.

) That is $bA = \underline{a} \supset uA$ where u is a unit mod $\underline{c}$. Therefore it is ufficient to prove theorem 12.11 for the case $P = A \oplus P'$, $_P = (b, a')$ and $bA + \underline{a} + \underline{c}A = A$. If d = dim m-spec(R) = 0 then = Jacobson radical of R. Hence $bA + \underline{a} = A$ by Nakayama's Lemma. ut $bA \subset O_P(a_P)$. Therefore $O_P(a_P) + \underline{a} = A$ and $f: Q \longrightarrow P$ the zero unction works. Case d = 0 of the theorem is done.

We will finish the proof by induction on d. First we need two emmas.

Lemma 12.9. Let R be a commutative ring, let $P_1, \ldots, P_n$ be finite collection of prime ideals, and let I be an ideal. If $\subset P_1 \cup \ldots \cup P_n$, then $I \subset P_i$ for some i.

Proof. This is obvious for n = 1. We can assume that I is ot contained in any n - 1 of the P_i. Pick $a_i \in (P_1 \cup \ldots \cup \hat{P}_i \cup \ldots \cup P_n) \cap I$, here $\hat{P}_i$ indicates that P_i is omitted from the union. Then $a_i \in P_i$ nd $a_i \notin P_j$ for $i \neq j$. Let $b_i = \prod_{j \neq i} a_j$. Then $b_i \notin P_i$ since P_i is rime and $b_i \in P_j$ for $j \neq i$. Let $b = \sum b_i$. Then $b \notin P_j$ for all j ut $b \in I$. This is a contradiction. Done with lemma 12.9.

Lemma 12.10. Let R be a commutative ring, let $P_1, \ldots, P_n$ be rime ideals, and let M be a finitely generated R module. If $/P_iM = 0$ for $i = 1, \ldots, n$, then there exists $t \in R$ such that $tM = 0$ nd $t \notin P_1 \cup \ldots \cup P_n$.

<u>Proof</u>. $P_{i_{P_i}} M_{P_i} = M_{P_i}$. Therefore $M_{P_i} = 0$ by Nakayama's Lemma. Hence, there exists $s_i \notin R - P_i$ such that $s_i M = 0$. Let I be the annihilator of M. Then $s_i \in I$. Therefore $I \not\subset P_i$ for all i. Therefore $I \not\subset P_1 \cup \ldots \cup P_n$ by lemma 12.9. Done with lemma 12.10.

By 3) we can assume that $bA + \underline{a} + \underline{c}A = A$. Let $M = A/(bA + \underline{a})$. Then $\underline{c}M = M$. Hence, $m_i M = M$ since $m_i \supset \underline{c}$. Therefore, by lemma 12.10, there exists $t \notin m_i$ such that $tM = 0$. That is $tA \subset bA + \underline{a}$. We try working mod t. Let $\bar{R} = R/(t)$. Then m-spec($\bar{R}$) = $\{m \in$ m-spec(R)$| m \supset (t)\}$ = = F(t). X = m-spec(R) = $X_1 \cup \ldots \cup X_n$ and $m_i \in X_i$. Therefore dim m-spec($\bar{R}$) < d.

We have that $O_Q(a_Q) + O(a') + bA + \underline{a} = A$ where $a_P = (a', b)$ since $P = P' \oplus A$. $\bar{a}_Q \in \bar{Q}$, $\overline{a'} \in \overline{P'}$, and $\overline{bA + \underline{a}}$ is a right ideal of $\bar{A}$. Therefore the inductive hypothesis applies since dim m-spec($\bar{R}$) is at least one smaller, f-rank$_m P' \geq$ f-rank$_m P - 1$, and f-rank on $\overline{P'} \geq$ f-rank P. Therefore there exists $\bar{f}: \bar{Q} \longrightarrow \bar{P}$ such that $O_{P'}(\overline{a'} + \bar{f}(\bar{a}_Q)) + \overline{bA + \underline{a}} = \bar{A}$. Q is projective so $\bar{f}$ lifts to f. Replace a' by $a' + f(a_Q)$. Then $O_{\overline{P'}}(\overline{a'}) + \overline{bA + \underline{a}} = \overline{O_{P'}(a')} + \overline{bA + \underline{a}} = A$. Hence $O_{P'}(a') + bA + \underline{a} + tA = A$. But $tA \subset bA + \underline{a}$. Therefore $O_{P'}(a') + bA + \underline{a} = O_P(a_P) + \underline{a} = A$. We need a map f: Q $\longrightarrow$ P such that $O_P(a_P + f(a_Q)) + \underline{a} = A$. Take f = 0. Done with theorem 12.7.

<u>Remark</u>. This also finishes the proof of theorem 12.3, corollary 12.4, theorem 12.5, and corollary 12.6.

<u>Application</u>. A and R as above. If [P] = [Q] in $K_0(A)$. Then $P \oplus A^r$ is isomorphic to $Q \oplus A^r$ for some r. If f-rank $P \geq d + 2$, P is isomorphic to Q. In particular if G is a finite group and

$A = Z[G]$, then P a projective A module implies $P = A^r \oplus I$ by Serre's theorem. I is a projective ideal and $K_0(A) = Z \oplus \tilde{K}_0(A)$ where $\tilde{K}_0(A)$ is finite. By cancellation $A^r \oplus I$ is isomorphic to $A^r \oplus J$ where I and J are projective ideals if and only if $A \oplus I$ is isomorphic to $A \oplus J$.

Chapter 13. $K_1(\underline{A})$

Let $\underline{A}$ be a full subcategory of an abelian category which contains 0, is closed under $\oplus$, and is equivalent to a small category.

Recall that $\underline{A}[x, x^{-1}]$ is the category whose objects are pairs (A, f) where $A \in \underline{A}$ and $f: A \to A$ is an iso and whose maps $(A, f) \to (B, g)$ is a map $h: A \to B$ such that

$$\begin{array}{ccc} A & \xrightarrow{f} & A \\ {\scriptstyle h}\downarrow & & \downarrow{\scriptstyle h} \\ B & \xrightarrow{g} & B \end{array}$$

commutes.

A sequence $0 \to (A', f') \to (A, f) \to (A'', f'') \to 0$ is exact in $\underline{A}[x, x^{-1}]$ if and only if $0 \to A' \to A \to A'' \to 0$ is exact in A.

<u>Definition</u>. $K_1(\underline{A})$ is the abelian group whose generators are $[(A, f)] \in \text{obj } \underline{A}[x, x^{-1}]$ and relations

1) If $0 \to (A', f') \to (A, f) \to (A'', f'') \to 0$ is exact in $\underline{A}[x, x^{-1}]$, then $[(A, f)] = [(A', f')] + [(A'', f'')]$ and

2) $[(A, fg)] = [(A, f)] + [(A, g)]$.

If $T: \underline{A} \to \underline{B}$ is a covariant functor preserving short exact sequences, it induces a map $K_1(T): K_1(\underline{A}) \to K_1(\underline{B})$ by $[(A, f)] \rightsquigarrow [(TA, Tf)]$.

Definition. If R is a ring, $K_1(R) = K_1$ (category of finitely generated projective modules).

If $f: R \to R'$ is a ring homomorphism, the functor $T(M) = R' \otimes_R M$ preserves short exact sequences of projective modules all of which split in $\underline{A}$. Therefore, K_1 is a functor on the category of rings and ring homomorphisms by $K_1(f) = K_1(T)$.

Remarks. 1) In topology $K^{-1}(X) = K^0(SX)$ where S is the suspension.

2) If R is a field, then $K_1(R) = R - \{0\} = R^*$ since $(P, f) \rightsquigarrow \det(f)$ is universal for the above conditions. Thus, K_1 can be thought of as a generalized determinant.

Proposition 13.1. Let R be a ring and let $\underline{A}$ be the category of finitely generated projective R modules and $\underline{F}$ be the category of finitely generated free R modules. Then $i: K_1(\underline{F}) \to K_1(R)$ is an iso where the map is the map induced from the inclusion of $\underline{F}$ into $\underline{A}$.

Proof. Let $(P, f) \in K_1(R)$. Then there exists Q such that $P \oplus Q = F$ is a finitely generated free module. Let $g: F \to F$ be given by $g(p, q) = (f(p), q)$. Define $j: K_1(R) \to K_1(\underline{F})$ by $j([(P, f)]) = [(F, g)]$. In K_1 (any category), $[(A, 1_A)] = [(A, 1_A \circ 1_A)] = [(A, 1_A)] + [(A, 1_A)]$. Therefore $[(A, 1_A)] = 0$. Therefore $ij([(P, f)]) = [(F, g)] = [(P \oplus Q, f \oplus 1_Q)]$ $= [(P, f)] + [Q, 1_Q] = [(P, f)]$. Therefore $ij = 1_{K_1(R)}$. $ji = 1_{K_1(\underline{F})}$ is clear. Hence we only need to show that j is well defined. Suppose $P \oplus Q' = F'$ is a finitely generated free module and $g' = (f \oplus 1_{Q'})$. Then in $K_1(\underline{F})$, $[(P \oplus Q \oplus P \oplus Q', f \oplus 1 \oplus 1 \oplus 1)] = [(P \oplus Q \oplus P \oplus Q', 1 \oplus 1 \oplus f \oplus 1)]$ since the objects involved are isomorphic. Therefore $[(P \oplus Q, f \oplus 1)] \oplus 0 = 0 + [(P \oplus Q', f \oplus 1)]$. Hence j is well defined. Done.

$$\begin{aligned} \mathrm{Aut}(R^n) &= \text{the group of } n \times n \text{ invertible matrices} \\ &= GL(n, R). \end{aligned}$$

We can identify $f \in GL(n, R)$ with $(R^n, f) \in \underline{F}[X, X^{-1}]$. This gives us a map $GL(n, R) \longrightarrow K_1(\underline{F})$ which is clearly a homomorphism of groups. We can map $GL(n, R) \longrightarrow GL(n + 1, R)$ by sending the matrix (X) to $\begin{pmatrix} X & \begin{matrix} 0 \\ \vdots \\ 0 \end{matrix} \\ 0 \dots 0 & 1 \end{pmatrix}$. Say $f \in GL(n, R)$ goes to $f' = f \oplus 1_R$. Then $[(R^{n+1}, f')] = [(R^n, f)] + [(R, 1_R)] = [(R^n, f)]$. Hence the map $GL(n, R) \longrightarrow GL(n + 1, R)$ commutes with the map to $K_1(\underline{F})$.

<u>Definition</u>. Let <u>$GL(R)$</u> $= \varinjlim GL(n, R)$.

The above discussion gives us a map $GL(R) \longrightarrow K_1(R)$ which is onto by proposition 13.1. We can ask what the kernel is. Since $K_1(R)$ is an abelian group the kernel must contain $[GL(R), GL(R)]$.

<u>Definitions</u>. $M \in GL(n, R)$ is an <u>elementary matrix</u> if $M = I + ae_{ij}$ where $i \neq j$ and e_{ij} is the matrix with 1 in the i-th row j-th column and 0 elsewhere. <u>$E(n, R)$</u> is the subgroup generated by all $n \times n$ elementary matrices. <u>$E(R)$</u> $= \varinjlim E(n, R)$ where $\varinjlim$ is taken over the maps $GL(n, R) \longrightarrow GL(n + 1, R)$ (which do send elementary matrices to elementary matrices).

<u>Theorem 13.2</u>. Let R be any ring. Then $E(R) = [GL(R), GL(R)]$.

Remark. The inclusion $E(R) \supset [GL(R), GL(R)]$ was proved by J. H. C. Whitehead. I. Kaplansky then remarked that this inclusion was, in fact, an equality. The reverse inclusion holds even in the finite case if $n \geq 3$.

Proof. Let e_{ij} be the matrices as above. Then

$$e_{ij}e_{kl} = \begin{cases} 0 \text{ if } j \neq k \\ e_{il} \text{ if } j = k. \end{cases}$$

To show $E(R) \subset [GL(R), GL(R)]$ we will show that we can obtain all the generators $1 + ae_{ij}$. Pick $k \neq i$ or j. Then we claim $1 + ae_{ij} = [1 + ae_{ik}, 1 + e_{kj}]$. It is clear that $(1 + ae_{ij})^{-1} = (1 - ae_{ij})$ if $i \neq j$. Hence

$$\begin{aligned} [1 + ae_{ik}, 1 + e_{kj}] &= (1 + ae_{ik})(1 + e_{kj})(1 - ae_{ik})(1 - e_{kj}) \\ &= (1 + ae_{ik} + e_{kj} + ae_{ij})(1 - ae_{ik})(1 - e_{kj}) \\ &= (1 + ae_{ik} + e_{kj} + ae_{ij} - ae_{ik})(1 - e_{kj}) \\ &= (1 + e_{kj} + ae_{ij})(1 - e_{kj}) \\ &= (1 + e_{kj} + ae_{ij} - e_{kj}) \\ &= (1 + ae_{ij}). \end{aligned}$$

Therefore, $E(R) \subset [GL(R), GL(R)]$.

We can think of the $2n \times 2n$ matrices over R as 2×2 matrices over the $n \times n$ matrices over R. Then $\begin{pmatrix} I & * \\ 0 & I \end{pmatrix}$ is the product of matrices in $E(2n, R)$ where * is any $n \times n$ matrix over R. This is proved by reducing $\begin{pmatrix} I & * \\ 0 & I \end{pmatrix}$ to $\begin{pmatrix} I & 0 \\ 0 & I \end{pmatrix}$ by a sequence of row and column operations.

Lemma 13.3. Let $A \in GL(n, R)$, then $\begin{pmatrix} A & 0 \\ 0 & A^{-1} \end{pmatrix} \in E(2n, R)$.

Proof. We exhibit a sequence of steps each of which is given by multiply by an element of E(2n, R) and the last position is the identity. Reversing the procedure gives $\begin{pmatrix} A & 0 \\ 0 & A^{-1} \end{pmatrix}$.

$$\begin{pmatrix} A & 0 \\ 0 & A^{-1} \end{pmatrix} \longrightarrow \begin{pmatrix} A & I \\ 0 & A^{-1} \end{pmatrix} \longrightarrow \begin{pmatrix} 0 & I \\ -I & A^{-1} \end{pmatrix} \longrightarrow \begin{pmatrix} 0 & I \\ -I & 0 \end{pmatrix} \longrightarrow \begin{pmatrix} I & 0 \\ 0 & I \end{pmatrix} .$$

Done with lemma 13.3.

Now we prove $E(R) \supset [GL(R), GL(R)]$. We only need to obtain all the generators of [GL(R), GL(R)]. Pick A and B $\in$ GL(R). Find n big enough so that A and B can be represented in GL(n, R). Then $\begin{pmatrix} [A, G] & 0 \\ 0 & I \end{pmatrix}$ is in GL(2n, R). We claim it is in E(2n, R).

$$\begin{pmatrix} ABA^{-1}B^{-1} & 0 \\ 0 & I \end{pmatrix} = \begin{pmatrix} AB & 0 \\ 0 & B^{-1}A^{-1} \end{pmatrix} \begin{pmatrix} A^{-1} & 0 \\ 0 & A \end{pmatrix} \begin{pmatrix} B^{-1} & 0 \\ 0 & B \end{pmatrix} .$$

But each matrix on the right hand side is in E(2n, R) by lemma 13.3 Therefore, $\begin{pmatrix} [A, B] & 0 \\ 0 & I \end{pmatrix} \in E(2n, R)$ and $[GL(R), GL(R)] \in E(2n, R)$.

Done with theorem 13.2.

Theorem 13.4. Let R be any ring. Then the map $GL(R)/E(R) = GL(R)/[GL(R), GL(R)] \longrightarrow K_1(R) = K_1(\underline{F})$ is an iso.

Proof. We construct a map $j: K_1(\underline{F}) \longrightarrow GL(R)/E(R)$ which is an inverse. Let $[(F, f)] \in K_1(\underline{F})$. Pick a basis $e_1, \ldots, e_n$ for F. Then f is represented by a matrix $A \in GL(n, R)$. Define $j([(F, f)]) = [A]$ in GL(R)/E(R). If j is well defined, then j is clearly the inverse. Let $(F, f) \in \underline{F}[X, X^{-1}]$. Suppose we picked basis $e_1, \ldots, e_n$ for F and f was represented by A with respect to that basis. Let $e_1', \ldots, e_m'$ be another basis for A possibly with a different number of elements m such that f is represented by A' with respect to this basis. Then in $\underline{F}[X, X^{-1}]$ there is an an iso h: $(F \oplus F, f \oplus 1) \to (F \oplus F, 1 \oplus f)$. Consider the two $n + m \times n + m$ matrices $\begin{pmatrix} A & 0 \\ 0 & I \end{pmatrix}$ and $\begin{pmatrix} I & 0 \\ 0 & A' \end{pmatrix}$ which are obtained from the basis $e_1, \ldots, e_n, e'_1, \ldots, e'_m$. The iso h gives a matrix $C \in GL(n + m, R)$ such that $\begin{pmatrix} A & 0 \\ 0 & I \end{pmatrix} = C\begin{pmatrix} I & 0 \\ 0 & A' \end{pmatrix}C^{-1}$. Hence, $\begin{pmatrix} A & 0 \\ 0 & I \end{pmatrix}$ and $\begin{pmatrix} I & 0 \\ 0 & A' \end{pmatrix}$ give the same element in $GL(R)/E(R)$. But these have the same images in $GL(R)/E(R)$ as A and A'.

To complete the proof that j is well defined we need to show that j preserves relations. The relations $[(F, fg)] = [(F, f)] + [(F, g)]$ are immediate. Pick a basis $e_1, \ldots, e_n$ for F. Then if A represents f and B represents g then AB clearly represents fg. Let

$0 \to (F', f') \to (F, f) \to (F'', f'') \to 0$ be an exact sequence in $\underline{F}[X, X^{-1}]$. Then

$0 \to F' \to F \to F'' \to 0$ is exact in $\underline{F}$.

Pick a basis $e'_1, \ldots, e'_n$ for F'. Extend it to a basis $e'_1, \ldots, e'_n, e''_1, \ldots, e''_m$ for F by choosing the e''_i so that their images in F" form a basis for F". Let A represent f' and A" represent f". Then $A = \begin{pmatrix} A' & B \\ 0 & A'' \end{pmatrix}$ represents f where B is an $n \times m$ matrix. But then

$$A = \begin{pmatrix} A' & B \\ 0 & A'' \end{pmatrix} = \begin{pmatrix} A' & 0 \\ 0 & I \end{pmatrix} \begin{pmatrix} I & 0 \\ 0 & A'' \end{pmatrix} \begin{pmatrix} I & B \\ 0 & I \end{pmatrix},$$

$\begin{pmatrix} I & B \\ 0 & I \end{pmatrix}$ is in $E(R)$, and $\begin{pmatrix} I & 0 \\ 0 & A'' \end{pmatrix}$ is equivalent mod $E(R)$ to $\begin{pmatrix} A'' & 0 \\ 0 & I \end{pmatrix}$. Hence

$$j([(F, f)]) = j([(F', f')]) + j([(F'', f'')]).$$

Therefore both relations are preserved and j is well defined. Done.

Remark. If R^* is the group of units of R, we always have a map $R^* \to K_1(R)$ by sending $a \in R^*$ to $\begin{pmatrix} a & 0 \\ 0 & I \end{pmatrix} \in GL(R)$ and then to $K_1(R)$ by the isomorphism above. If R is commutative, then the determinant map commutes with the inclusions $GL(n, R) \to GL(n + 1, R)$. Hence we get a map det: $GL(R) \to R^*$. Since R^* is commutative det factors through $K_1(R)$. The composition $R^* \to K_1(R) \to R^*$ is clearly the identity. Hence if R is commutative, $K_1(R) = R^* \oplus SK_1(R)$ where $SK_1(R) = \ker(K_1(R) \to R^*)$. Let $SL(n, R) = \ker(GL(n, R) \xrightarrow{\det} R^*)$ and let $SL(R) = \varinjlim SL(n, R)$. We have $E(R) \subset SL(R) \subset GL(R)$ and $SL(R)/E(R)$ is isomorphic to $SK_1(R)$ and $GL(R)/SL(R)$ is isomorphic to R^*. If R is a field, then $SL(n, R) = E(n, R)$ for each n and hence $SK_1(R) = 0$ and $K_1(R) = R^* = R - \{0\}$.

By theorem 13.4 we have maps

$$GL(n, R)/E(n, R) \to GL(n + 1, R)/E(n + 1, R) \to \cdots \to GL(R)/E(R)$$

which is isomorphic to $K_1(R)$. All the objects on the left are just left cosets since $E(n, R)$ can fail to be normal in $GL(n, R)$. We can ask if one of the maps $GL(n, R)/E(n, R) \to K_1(R)$ is actually onto or an iso. We can also ask if any of the $GL(n, R)/E(n, R)$ are actually groups or abelian groups. Our next goal is to give a partial answer to this.

Definition. n defines a stable range for $GL(R)$ if whenever $r > n$ and $(a_1, \ldots, a_r)$ is a unimodular row, then there exists $b_1, \ldots, b_{r-1} \in R$ such that $(a_1 + a_r b_1, \ldots, a_{r-1} + a_r b_{r-1})$ is a unimodular row.

Remark. Theorem 12.3 proved that if R is a commutative ring such that m-spec(R) is a noetherian space of dim $d < \infty$ and if A is an R algebra which is finitely generated as an R module, then d + 1 defines a stable range for GL(A).

Theorem 13.5. If n defines a stable range for R then

1) $GL(m, R)/E(m, R) \to GL(R)/E(R)$ is onto if $m \geq n$,

2) $E(r, R)$ is a normal subgroup of $GL(r, R)$ if $r \geq n + 1$, and

3) $GL(r, R)/E(r, R)$ is an abelian group if $r \geq 2n$.

Remark. Bass, Milnor, and Serre have proved for R commutative $GL(m, R)/E(m, R) \xrightarrow{\approx} GL(R)/E(R)$ whenever $m \geq n + 2$ where n defines a stable range for GL(R). The proof is by constructing an inverse. We will omit the proof in these notes. Presumably their proof extends to the noncommutative case. It is also conjectured that the map $GL(n + 1, R)/E(n + 1, R) \to GL(n + 2, R)/E(n + 2, R)$ is an iso when n defines a stable range for GL(R).

Lemma 13.6. If n defines a stable range for GL(R), $r > n$, and $(a_1, \ldots, a_r)$ is a unimodular row, then there exists an $A \in E(r, R)$ such that $(a_1, \ldots, a_r)A = (1, 0, \ldots, 0)$.

Proof. Since n defines a stable range for GL(R), there exist $b_1, \ldots, b_{r-1}$ such that $(a_1 + a_r b_r, \ldots, a_{r-1} + a_r b_{r-1})$ is unimodular. The transformation $(a_1, \ldots, a_r) \to (a_1 + a_r b_1, \ldots, a_{r-1} + a_r b_{r-1}, a_r)$ is clearly obtainable by an elementary matrix. Hence we can assume $(a_1, \ldots, a_{r-1})$ is unimodular. Then there exist c_i such that $1 = a_1 c_1 + \ldots + a_{r-1} c_{r-1}$. Therefore, $-a_r + 1 = a_1 c_1(-a_r + 1) + \ldots + a_{r-1} c_{r-1}(-a_r + 1)$. Therefore the transformation $(a_1, \ldots, a_r) \to (a_1, \ldots, a_{r-1}, 1)$ is obtainable by an elementary matrix. Finally we use the 1 to kill all the other a_i and then switch 1 to first position. That is we send

$(a_1, \ldots, a_{r-1}, 1) \longrightarrow (0, \ldots, 0, 1) \longrightarrow (1, 0, \ldots, 0, 1) \longrightarrow$ $(1, 0, \ldots, 0)$. Done with lemma 13.6.

Corollary 13.7. R and n as above. If $r \geq n$, then $GL(r, R)E(r + 1, R) = GL(r + 1, R)$.

Proof. Let $A \in GL(r + 1, R)$. Then the bottom row is unimodular. Therefore, there exists $E \in E(r + 1, R)$ such that the bottom row of AE is $(0, \ldots, 0, 1)$. Say $AE = \begin{pmatrix} A' & f \\ 0 & 1 \end{pmatrix}$ where $A' \in GL(r, R)$ and f is a column of length r. Any matrix of the form $\begin{pmatrix} I & f \\ 0 & 1 \end{pmatrix}$ where I is $r \times r$ identity and f is a column of length r is in $E(r + 1, R)$ since $\begin{pmatrix} I & f \\ 0 & 1 \end{pmatrix}\begin{pmatrix} I & f' \\ 0 & 1 \end{pmatrix} = \begin{pmatrix} I & f + f' \\ 0 & 1 \end{pmatrix}$ and we could pick up one non zero piece of f at a time. Let $g = -A'^{-1}f$

$$AE\begin{pmatrix} I & g \\ 0 & 1 \end{pmatrix} = \begin{pmatrix} A' & f \\ 0 & 1 \end{pmatrix}\begin{pmatrix} I & g \\ 0 & 1 \end{pmatrix} = \begin{pmatrix} A' & A'g + f \\ 0 & 1 \end{pmatrix} = \begin{pmatrix} A' & -f + f \\ 0 & 1 \end{pmatrix} = \begin{pmatrix} A' & 0 \\ 0 & 1 \end{pmatrix}.$$

Therefore $AE\begin{pmatrix} 1 & g \\ 0 & 1 \end{pmatrix} \in GL(r, R)$ and $GL(r, R)E(r + 1, R) = GL(r + 1, R)$

Done with corollary 13.7.

1) of theorem 13.5 is clear from corollary 13.7.

2) We want to show that $E(r + 1, R)$ is normal in $GL(r + 1, R)$. By corollary 13.7 it is enough to show that the normalizer of $E(r + 1, R)$ in $GL(r + 1, R)$ contains $E(r + 1, R)$ and $GL(r, R)$. It certainly contains $E(r + 1, R)$. Therefore, we only have to show that if $E \in E(r + 1, R)$ and $A \in GL(r, R)$ then $\begin{pmatrix} A & 0 \\ 0 & 1 \end{pmatrix}E\begin{pmatrix} A^{-1} & 0 \\ 0 & 1 \end{pmatrix} = E(n + 1, R)$. It will clearly suffice to do the case where E itself is elementary. We divide this case into three cases:

a) $E = \begin{pmatrix} I & a \\ 0 & 1 \end{pmatrix}$ where a is a column vector of length r,

b) $E = \begin{pmatrix} I & 0 \\ b & 1 \end{pmatrix}$ where b is a row vector of length r, and

c) $E = \begin{pmatrix} E' & 0 \\ 0 & 1 \end{pmatrix}$ where $E' \in E(r, R)$.

In case a) we have $\begin{pmatrix} A & 0 \\ 0 & 1 \end{pmatrix}\begin{pmatrix} I & a \\ 0 & 1 \end{pmatrix}\begin{pmatrix} A^{-1} & 0 \\ 0 & 1 \end{pmatrix} = \begin{pmatrix} I & Aa \\ 0 & 1 \end{pmatrix}$ which is a product of elementary matrices.

Case b) is exactly the same.

In case c) let $E' = I + qe_{ij}$ where $i, j \leq r$ and $i \neq j$. Then the result of conjugating is $\begin{pmatrix} AE'A^{-1} & 0 \\ 0 & 1 \end{pmatrix}$.

$AE'A^{-1} = I + Aqe_{ij}A^{-1}$. Now $A\,qe_{ij}A^{-1}$ has entry $a_{\gamma i}qb_{jk}$ in the γk position where $A = (a_{ij})$ and $A^{-1} = (b_{ij})$. Let a be the ith column of A and b be the j-th row of $B = A^{-1}$. Then aqb is an $r \times r$ matrix and ba is a $|x|$ matrix. $ba = \sum b_{j\gamma}a_{\gamma i}$ = ji-th entry of $A^{-1}A = 0$ since $i \neq j$. Therefore we have a matrix $\begin{pmatrix} I + aqb & 0 \\ 0 & 1 \end{pmatrix}$ where a is a column of length r and, b is a row, q is 1×1, $ba = 0$, and $I + aqb$ is invertible. From this we want to conclude that $\begin{pmatrix} I + aqb & 0 \\ 0 & 1 \end{pmatrix} \in E(r + 1, R)$. Let g be any row of length r. Then

$$\begin{pmatrix} I & 0 \\ g & 1 \end{pmatrix}\begin{pmatrix} I + aqb & 0 \\ 0 & 1 \end{pmatrix} = \begin{pmatrix} I + aqb & 0 \\ g(I + aqb) & 1 \end{pmatrix}.$$

Hence by picking g properly we can obtain any row for $g(I + aqb)$ since $I + aqb$ is invertible. Pick g so that the result is

$$\begin{pmatrix} I + aqb & 0 \\ qb & 1 \end{pmatrix}$$

Now $\begin{pmatrix} I & -a \\ 0 & 1 \end{pmatrix}\begin{pmatrix} I + aqb & 0 \\ qb & 1 \end{pmatrix} = \begin{pmatrix} I & -a \\ qb & 1 \end{pmatrix}$ and

$$\begin{pmatrix} I & -a \\ qb & 1 \end{pmatrix}\begin{pmatrix} I & a \\ 0 & 1 \end{pmatrix} = \begin{pmatrix} I & a - a \\ qb & qba + 1 \end{pmatrix} = \begin{pmatrix} I & 0 \\ qb & 1 \end{pmatrix}.$$

But $\begin{pmatrix} I & 0 \\ qb & 1 \end{pmatrix}$ is clearly a product of elementary matrices. Therefore for all possibilities of generators of E(r + 1, R) conjugating by GL(r, R) sends them into E(r + 1, R). Hence, GL(r, R) normalizes E(r + 1, R) and E(r + 1, R) is normal in GL(r + 1, R) and 2) of theorem 13.5 is done.

3) By part 2) GL(r, R)/E(r, R) is a group since $r \geq 2n$ and 0 cannot be a stable range. The proof of theorem 13.2 actually shows that $E(2r, R) \supset [GL(r, R), GL(r, R)]$ for every r. By part 1) the map $GL(n, R) \to GL(r, R)/E(r, R)$ is onto. Pick $x, y \in GL(r, R)/E(r, R)$ and preimages $x', y' \in GL(n, R)$. Then $[x', y'] \in E(2n, R) \subset E(r, R)$. But $[x', y']$ is a pre image of $[x, y]$. Therefore, $[x, y] \in E(2n, R) \subset E(r, R)$. Therefore, GL(r, R)/E(r, R) is an abelian group. Done with theorem 13.5.

<u>Example</u>. Let R be a Euclidean ring. Then given $x, y \in R$ there is a q such that $y = xq + r$ and $|r| < |x|$ or $r = 0$. If (a, b) is a unimodular row then $aR + bR = R$. We can replace (a, b) by (a, r) where $b = aq + r$ and $|r| < |b|$. This continues until we get (a', b') and $|b| < |a|$ and b divides a. Then b is a unit. $(a', b') \to (1, b')$ is elementary. Finally $(1, b') \to (1, 0)$ is elementary. Therefore, in a sequence of steps given by elementary matrices we can send any unimodular row (a, b) to (1, 0). Clearly, the same thing works for rows of greater length. Therefore, n = 1 defines a stable range for R. Therefore, $GL(1, R) = R^*$ maps onto $K_1(R)$. This is split by the determinant map. Hence, det: $K_1(R) \to R^*$ is an iso and $SK_1(R) = 0$. It is unknown if this is true for arbitrary commutative principal ideal domains. Bass, Milnor, and Serre have shown if R is the ring of algebraic number field then $SK_1(R) = 0$. The following example of Milnor shows that there is a Dedekind ring with

$SK_1(R) \neq 0$. Let $R = \mathbb{R}[x, y]/(x^2 + y^2 - 1)$. Regarding R as a ring of functions on the circle gives a map from SL(n, R) to the function space $SL(n, \mathbb{R})^{S_1}$ and thus a map $SL(n, R) \longrightarrow \pi_1(SL(n, \mathbb{R}))$. These maps are compatible and give a map $SK_1(R) \longrightarrow \pi_1(SL(\mathbb{R})) = Z/2Z$. The matrix $\begin{pmatrix} x & y \\ -y & x \end{pmatrix}$ maps into a generator of $\pi_1(SL(\mathbb{R}))$ so $SK_1(R) \neq 0$. It can be shown that $SK_1(R) = Z/2Z$.

Chapter 14. $K_2(R)$

In this chapter we present Milnor's definition of $K_2(R)$.

Let R be a ring. Then E(R) is generated by $I + ae_{ij}$ where $a \in R$ and $i \neq j$. These generators have relations

$(I + re_{ij})(I + se_{ij}) = (I + (r + s)e_{ij})$ and

$$[I + se_{ij}, I + te_{k\ell}] = \begin{cases} I + ste_{i\ell} & \text{if } j = k \text{ and } i \neq \ell \\ I & \text{if } j \neq k \text{ and } i \neq \ell \\ I - tse_{kj} & \text{if } j \neq k \text{ and } i = \ell. \end{cases}$$

<u>Remarks</u>. These relations are a special case of those studied by Chevalley and Steinberg for algebraic groups over fields.

<u>Definition</u>. Let R be a ring. Then the Steinberg group of R, ST(R), is the (non abelian) group with generators $x_{ij}(t)$ for all positive integers $i \neq j$ and all $t \in R$, and with the relations $x_{ij}(s)x_{ij}(t) = x_{ij}(s + t)$ and

$$[x_{ij}(s), x_{k\ell}(t)] = \begin{cases} 1 & \text{if } i \neq \ell \text{ and } j \neq k \\ x_{i\ell} & \text{if } j = k \text{ and } i \neq \ell. \end{cases}$$

Remarks. Since in any group $[a, b]^{-1} = [b, a]$, the relation $[x_{ij}(s), x_{ki}(t)] = x(-ts)_{kj}$ is an immediate consequence. The function $f\colon ST(R) \to E(R)$ given by $f(x_{ij}(t)) = I + e_{ij}$ is a homomorphism of groups. It is clear that ST is a functor from rings to groups. We could have defined $ST(n, R)$, but it is not known if the map $ST(n, R) \to ST(R)$ is a mono.

Definition. $K_2(R)$ is the kernel of $f\colon ST(R) \to E(R)$.

Remark. We have an exact sequence of functors

$$0 \to K_2 \to ST \to GL \to K_1 \to 0.$$

Theorem 14.1. $K_2(R)$ is the center of $ST(R)$.

Proof. $Z(ST(R)) \subset K_2(R)$. Pick $a \in Z(ST(R))$. Since f is onto, $f(a) \in Z(E(R))$. Hence it is enough to show that $E(R)$ has a trivial center. We will show that $C_{GL(R)}(E(R))$, the centralizer of $E(R)$ in $GL(R)$, is the identity. Suppose

$$(\sum c_{ij}e_{ij})(1 + te_{uv}) = (1 + te_{uv})(\sum c_{ij}e_{ij}).$$

Then $(\sum c_{ij}e_{ij})(te_{uv}) = (te_{uv})(\sum c_{ij}e_{ij})$. Therefore,

$\sum_i c_{iu}te_{iv} = \sum_j tc_{vj}e_{uj}$. Therefore, if $i \neq u$, then $c_{iu}t = 0$, and if $v \neq j$, then $tc_{vj} = 0$. But this happens for all $t \in R$. Therefore, $c_{ij} = 0$ if $i \neq j$. $c_{uu}te_{uv} = tc_{vv}e_{uv}$ implies $c_{uu}t = tc_{vv}$. Therefore, $c_{uu} = c_{vv}$ and $(c_{ij}) = cI$ where c is in the center of R. But $(c_{ij}) \in GL(R)$. Therefore (c_{ij}) is eventually the identity matrix. Hence $c_{ii} = 1$ for all i and (c_{ij}) is the identity matrix. Therefore, $Z(ST(R)) \subset K_2(R)$.

Now to show $Z(ST(R)) \supset K_2(R)$. Let C_n be the subgroup of $ST(R)$ which is generated by all $x_{in}(t)$ where $i \neq n$ and $t \in R$. Let R_n be the subgroup of ST(R) generated by all $x_{nj}(t)$ where $j \neq n$ and $t \in R$. Then the basic relations show that C_n and R_n are commutative since each pair of generators commute. Any product of the form $x_{\,n}(t_1)x_{\,n}(t_2) \ldots x_{\,n}(t_m)$ is just $x_{1n}(\sum_1^m t_j)$. Therefore any product in C_n can be written as $x_{1n}(t_1)x_{2n}(t_2) \ldots x_{mn}(t_m) \ldots\ldots$ where $t_v = 0$ for all v large enough.

Now $f(x_{1n}(t_1) \ldots x_{mn}(t_m) \ldots) = 1 + t_1 e_{\,n} + t_2 e_{2n} + \ldots .$ That is

$$\begin{pmatrix} 1 & & & t_1 & & \\ & \ddots & & \vdots & & \\ & & & 1 & & \\ & & & t_n & \ddots & \\ & & & \vdots & & \\ & & & & & 1 \end{pmatrix} = f(x_{1n}(t_1) \ldots x_{mn}(t_m) \ldots\ldots)$$

But we can obviously recover $x_{1n}(t_1) \ldots x_{mn}(t_m)$ from it. Hence f is one-one on C_n. By symmetry f is one-one on R_n.

If $p \neq n$, then $x_{pq}(t) \in N_{ST(R)}(C_n)$, the normalizer in ST(R) of C_n because by the basic relation

$$x_{pq}(t)x_{in}(s)x_{pq}(t)^{-1} = \begin{cases} x_{in}(s) \text{ if } q \neq i \\ x_{pn}(ts)x_{in}(s) \text{ if } q = i. \end{cases}$$

Similarly $x_{p,q}(t) \in N_{ST(R)}(R_n)$ if $q \neq n$.

Let $a \in ST(R)$ with $f(a) = 1$, i.e., $a \in K_2(R)$. We want to prove that $a \in Z(ST(R))$. By the above $a = \prod x_{i_v j_v}(t_v)$, a finite product. Pick $n \neq$ any i_v or j_v. Then $a \in N_{ST(R)}(C_n)$ and $a \in N_{ST(R)}(R_n)$. Let

$x \in C_n$, then $f(axa^{-1}) = f(x)$ since $f(a) = 1$. But $a \in N_{ST(R)}(C_n)$ so $axa^{-1} \in C_n$. But f is one-one on C_n. Therefore $axa^{-1} = x$. Hence, $a \in C(C_n)$, the centralizer of C_n. Similarly, $a \in C(R_n)$. But C_n and R_n generate ST(R) for we have all $x_{in}(t)$ and $x_{jn}(t)$ but if i and j are $\neq$ n then

$$x_{ij}(t) = [x_{in}(t), x_{nj}(1)] \quad \text{which}$$

is certainly in the subgroup generated by C_n and R_n. Therefore, $a \in Z(ST(R))$.

Done.

Proposition 14.2. $[ST(R), ST(R)] = ST(R)$ and $[E(R), E(R)] = E(R)$.

Proof. Let $n \neq i$ or j. Then $x_{ij}(t) = [x_{in}(t), x_{nj}(1)]$. Hence $ST(R) = [ST(R), ST(R)]$. The second part follows since f is onto and the image of a perfect group is perfect.

Done.

C. Moore has shown that one can ape a good deal of the classical topological theory of covering groups in abstract group theory. The analogue of a connected group is a perfect group and that of a covering is a central extension. Kervaire and Steinberg have shown that ST(R) is simply connected in this sense (see theorem 14.2 below). Therefore, we can regard E(R) as the connected component of GL(R) and ST(R) as the universal covering of E(R).

In terms of homology, $H_1(GL(R)) = K_1(R)$, $H_1(E(R)) = 0$, $H_2(E(R)) = K_2(R)$, and $H_1(ST(R)) = H_2(ST(R)) = 0$. This suggests trying to define $K_3(R)$ as $H_3(ST(R))$ and more generally to define $K_n(R)$ by killing homology groups. That is if there is a homomorphism $G \longrightarrow GL(R)$ with $H_i(G) = 0$ for $i = 1, \ldots, n-1$ and which is left universal for such maps and then setting $K_n(R) = H_n(G)$. However, there is no obvious reason why such a G should exist.

A number of other definitions of $K_n(R)$ have been proposed by various authors, but as yet no coherent theory of higher K's has appeared.

Theorem 14.2. Any central extension $0 \to A \to X \to ST(R) \to 0$ splits.

Proof. We need to lift the $x_{ij}(t) \in ST(R)$ to $f_{ij}(t) \in X$ such that the $f_{ij}(t)$ satisfy the relations for the $x_{ij}(t)$. Let $f_{ij}(t)$ and $f'_{ij}(t)$ be two liftings for the $x_{ij}(t)$. Then there exist $a_{ij}(t) \in A$ such that $f'_{ij}(t) = a_{ij}(t)\, f_{ij}(t)$. Let u and $v \in X$ and a and $b \in A$. Then $[u, v] = [au, bv]$ since $aubvu^{-1}a^{-1}v^{-1}b^{-1} = uvu^{-1}v^{-1}$ since A is central. Since $[x_{in}(t), x_{nj}(1)] = x_{ij}(t)$ if $n \neq i$ or j we can define $g_{ij}(t) = [f_{in}(t), f_{nj}(1)] = [f'_{in}(t), f'_{nj}(1)]$ when $n \neq i$ or j. The $g_{ij}(t)$ are independent of the lifting. We need to show that the $g_{ij}(t)$ satisfy the relations and are independent of n. $g_{ij}(t)$ clearly lifts $x_{ij}(t)$.

(1) Let $i \neq \ell$ and $j \neq k$. Then we will show that $[f_{ij}(s), f_{k\ell}(t)] = 1$ for any lifting. Define an automorphism b of X by $b(x) = f_{ij}(s)\, x\, f_{ij}(s)^{-1}$. Let $i \neq q$ and $j \neq p$. Then $[x_{ij}(s), x_{pq}(t)] = 1$ so $b(f_{pq}(t)) = a_{pq}(t)f_{pq}(t)$ where $a_{pq}(t) = [f_{ij}(s), f_{pq}(t)] \in A$. Pick $n \neq i, j, k$, or ℓ. Then $x_{k\ell}(t) = [x_{kn}(t), x_{n\ell}(1)]$ so $f_{k\ell}(t) = a[f_{kn}(t), f_{n\ell}(1)]$ where $a \in A$. Apply b to this. Then

$$\begin{aligned} b(f_{k\ell}(t)) &= b(a)[b(f_{kn}(t)), b(f_{n\ell}(1))] \\ &= a[a'f_{kn}(t), a''f_{n\ell}(1)] \text{ where } a' \text{ and } a'' \in A \\ &= a[f_{kn}(t), f_{n\ell}(1)] \text{ since A is central} \\ &= f_{k\ell}(t). \end{aligned}$$

hat is $b(f_{k\chi}(t)) = f_{ij}(s)f_{k\chi}(t)f_{ij}(s)^{-1} = f_{k\chi}(t)$ and

$f_{ij}(s), f_{k\chi}(t)] = 1.$

We claim that the $g_{ij}(t)$ satisfy the same relations as $x_{ij}(t)$

hen i and $j < n$. First we check that $g_{ij}(s + t) = g_{ij}(s)g_{ij}(t)$.

$$
\begin{aligned}
{}_{ij}(s + t) &= [f_{in}(s + t), f_{nj}(1)] \\
&= [af_{in}(s)f_{in}(t), f_{nj}(1)] \text{ where } a \in A \\
&= [f_{in}(s)f_{in}(t), f_{nj}(1)] \text{ since } a \in A \\
&= f_{in}(s)f_{in}(t)f_{nj}(1)f_{in}(t)^{-1}f_{in}(s)^{-1}f_{nj}(1)^{-1} \\
&= f_{in}(s)[f_{in}(t), f_{nj}(1)]f_{nj}(1)f_{in}^{-1}(s)f_{nj}(1)^{-1} \\
&= f_{in}(s)g_{ij}(t)f_{nj}(1)f_{in}(s)^{-1}f_{nj}(1)^{-1} \\
&= f_{in}(s)g_{ij}(t)f_{in}(s)^{-1}[f_{in}(s), f_{nj}(1)] \\
&= f_{in}(s)g_{ij}(t)f_{in}(s)^{-1}g_{ij}(s) \\
&= g_{ij}(t)g_{ij}(s) \text{ because } i \neq j,\ n \neq i \text{ so } [f_{in}(s), g_{ij}(t)] = 1 \text{ by (1)}
\end{aligned}
$$

For the same reason, $[g_{ij}(t), g_{ij}(s)] = 1$. Hence $g_{ij}(t)g_{ij}(s)$

$$
\begin{aligned}
&= g_{ij}(s)g_{ij}(t) \\
&= g_{ij}(s + t) \text{ as desired.}
\end{aligned}
$$

) Finally we need to check that $[g_{ij}(s), g_{jk}(t)] = g_{ik}(st)$ if $i \neq k$.

efine $b(x) = g_{ij}(s)x\, g_{ij}(s)^{-1}$. Then we want to prove that

$(g_{jk}(t)) = g_{ik}(st)g_{jk}(t)$. Now $g_{jk}(t) = [f_{jn}(t), f_{nk}(1)]$. Apply

to this and get $b(g_{jk}(t)) = [b(f_{jn}(t), b(f_{nk}(1))]$. Now

$$
\begin{aligned}
(x) = f_{ij}(s)x\, f_{ij}(s)^{-1} \text{ so } b(f_{jn}(t)) &= [f_{ij}(s), f_{jn}(t)]f_{jn}(t) \\
&= a\, f_{in}(st)f_{jn}(t) \text{ where } a \in A.
\end{aligned}
$$

And $b(f_{nk}(1)) = f_{ij}(s)f_{nk}(1)f_{ij}(s)^{-1}$

$= f_{nk}(1)$ since $i \neq k$ and $j < n$.

Therefore, $b(g_{jk}(t) = [a\ f_{in}(st)f_{jn}(t),\ f_{nk}(1)]$

$= [f_{in}(st)f_{jn}(t),\ f_{nk}(1)]$ since $a \in A$,

so $b(g_{jk}(t)) = f_{in}(st)f_{jn}(t)f_{nk}(1)f_{jn}(t)^{-1}f_{in}(st)^{-1}f_{nk}(1)^{-1}$

$= f_{in}(st)[f_{jn}(t),\ f_{nk}(1)]f_{nk}(1)f_{in}(st)^{-1}f_{nk}(1)^{-1}$

$= f_{in}(st)g_{jk}(t)f_{in}(st)^{-1}g_{ik}(st)$.

But $j < n$ and $i \neq k$ so $f_{in}(st)$ and $g_{jk}(t)$ commute by (1). Therefore $b(g_{jk}(t)) = g_{jk}(t)g_{ik}(st)$. But $i \neq k$ and $j \neq k$. Therefore $g_{jk}(t)g_{ik}(st) = g_{ik}(st)g_{jk}(t)$. Therefore, $b(g_{jk}(t)) = g_{ij}(s)g_{jk}(t)g_{ij}(s)^{-1}$

$= g_{ik}(st)g_{jk}(t)$,

or $[g_{ij}(s),\ g_{jk}(t)] = g_{ik}(st)$ when $i \neq k$ or j and i, j, and $k < n$.

Hence the commutator relations in 2) hold.

3) Now we want to check that the $g_{ij}(t)$ are independent of n. Let $h_{ij}(t) = [f_{im}(t),\ f_{mj}(1)]$ where $m > n$. Then the $h_{ij}(t)$ satisfy all the relations for i and $j < m$. Therefore

$h_{ij}(t) = [h_{in}(t),\ h_{nj}(1)]$ if $i \neq j$ and i and $j < n$

$= [af_{in}(t),\ a'f_{nj}(1)]$ where a and $a' \in A$

since $h_{\chi_r}(t)$ lifts $x_{\chi_r}(t)$

$= [f_{in}(t),\ f_{nj}(1)]$ since a and $a' \in A$

$= g_{ij}(t)$.

Therefore, the g_{ij} are independent of n so we can define $g_{ij}(t)$ for all i, j, t lifting $x_{ij}(t)$ and satisfying the same relations.

We get a homomorphism $ST(R) \to X$ which splits $X \to ST(R)$ by sending $x_{ij}(t)$ to $g_{ij}(t)$. Done with theorem 14.2.

Chapter 15. The Exact Sequence of K_i's.

In this chapter we define the relative K_i's and present an exact sequence for them.

Definitions. Let R be a ring and $\underline{A}$ a two sided ideal of R. $GL(R, \underline{A})$ is the kernel of $GL(R) \to GL(R/\underline{A})$. $ST(R, \underline{A})'$ is the kernel of $ST(R) \to ST(R/\underline{A})$. $E(R, \underline{A})$ is the kernel of $E(R) \to E(R/\underline{A})$. $K_2(R, \underline{A})'$ is the kernel of $ST(R, \underline{A})' \to GL(R, \underline{A})$. $K_1(R, \underline{A})$ is the cokernel of $ST(R, \underline{A}) \to GL(R, \underline{A})$. This can be summarized by the commutative diagram

$$\begin{array}{ccccccccc}
 & & 0 & & 0 & & 0 & & \\
 & & \downarrow & & \downarrow & & \downarrow & & \\
0 & \to & K_2(R, \underline{A})' & \to & K_2(R) & \to & K_2(R/\underline{A}) & & \\
 & & \downarrow & & \downarrow & & \downarrow & & \\
0 & \to & ST(R, \underline{A})' & \to & GL(R) & \to & ST(R/\underline{A}) & \to & 0 \\
 & & \downarrow & & \downarrow & & \downarrow & & \\
0 & \to & GL(R, \underline{A}) & \to & GL(R) & \to & GL(R/\underline{A}) & & \\
 & & \downarrow & & \downarrow & & \downarrow & & \\
 & & K_1(R, \underline{A}) & \to & K_1(R) & \to & K_1(R/\underline{A}) & & \\
 & & \downarrow & & \downarrow & & \downarrow & & \\
 & & 0 & & 0 & & 0 & &
\end{array}$$

where all rows and columns are exact.

It is clear that $ST(R) \to ST(R/\underline{A})$ is an epi since we obtain $ST(R/\underline{A})$ from $ST(R)$ by identifying $x_{ij}(s)$ and $x_{ij}(t)$ if $s \equiv t \bmod \underline{A}$. The kernel is generated normally by $x_{ij}(a)$ where $a \in \underline{A}$ since $x_{ij}(s) + x_{ij}(a) = x_{ij}(s + a)$.

Therefore, $ST(R, \underline{A})'$ is the normal closure in $ST(R)$ of the group generated by the $x_{ij}(a)$ for all $a \in \underline{A}$. $E(R, \underline{A})$ is the normal subgroup of $E(R)$ generated by all $I + ae_{ij}$ where $a \in \underline{A}$. $K_2(R, A)'$ is included in $K_2(R)$ by a simple diagram chase. Therefore, $K_2(R, \underline{A})'$ is abelian. We want to prove that $K_1(R, \underline{A})$ is abelian. First we must show that $E(R, \underline{A})$, the image of $ST(R, \underline{A})$, is normal in $GL(R, \underline{A})$. Second we must show that $E(R, \underline{A}) \supset [GL(R, \underline{A}), GL(R, \underline{A})]$.

<u>Theorem 15.1.</u> $E(R, \underline{A}) = [GL(R), GL(R, \underline{A})] = [E(R), E(R, \underline{A})]$.

<u>Corollary 15.2</u>. $E(R, \underline{A}) \supset [GL(R, \underline{A}), GL(R, \underline{A})]$, and, therefore, $K_1(R, \underline{A})$ is defined and abelian.

<u>Proof</u>. $I + ae_{ij} = [I + e_{ik}, I + ae_{kj}]$ if $k \neq i$ or j since $x_{ij}(a) = [x_{ik}(1), x_{kj}(a)]$ in $ST(R)$. Hence if $a \in \underline{A}$, then $I + ae_{ij} \in [E(R), E(R, \underline{A})]$. But $[E(R), E(R, \underline{A})]$ is a normal subgroup of $E(R)$. Therefore $E(R, \underline{A}) \subset [E(R), E(R, \underline{A})] \subset [GL(R), GL(R, \underline{A})]$. To prove $[GL(R), GL(R, \underline{A})] \subset E(R, \underline{A})$ we need to prove a lemma.

<u>Lemma 15.3</u>. Let $a \in GL(n, R)$ and $b \in GL(n, R, \underline{A})$, the kernel of $GL(n, R) \longrightarrow GL(n, R/\underline{A})$. Then

$$\begin{pmatrix} ab & 0 \\ 0 & 1 \end{pmatrix} = \begin{pmatrix} a & 0 \\ 0 & b \end{pmatrix} = \begin{pmatrix} ba & 0 \\ 0 & 1 \end{pmatrix} \quad \text{as}$$

left cosets of $E(2n, R, \underline{A})$.

<u>Proof</u>. The matrix $b - I$ goes to 0 over $R/\underline{A}$. Therefore, $b = I + q$ where q is a matrix with all its entries in $\underline{A}$.

Now $\begin{pmatrix} ba & 0 \\ 0 & 1 \end{pmatrix}\begin{pmatrix} 1 & (ba)^{-1}q \\ 0 & 1 \end{pmatrix}\begin{pmatrix} 1 & 0 \\ -a & 1 \end{pmatrix}\begin{pmatrix} 1 & -a^{-1}q \\ 0 & 1 \end{pmatrix}\begin{pmatrix} 1 & 0 \\ a & 1 \end{pmatrix}\begin{pmatrix} 1 & 0 \\ -b^{-1}qa & 1 \end{pmatrix} =$
$= \begin{pmatrix} a & 0 \\ 0 & b \end{pmatrix}$. But $\begin{pmatrix} 1 & (ba)^{-1}q \\ 0 & 1 \end{pmatrix}$, $\begin{pmatrix} 1 & -a^{-1}q \\ 0 & 1 \end{pmatrix}$, and $\begin{pmatrix} 1 & 0 \\ -b^{-1}qa & 1 \end{pmatrix} \in E(R, \underline{A})$.

$\begin{pmatrix} 1 & 0 \\ -a & 1 \end{pmatrix}$ and $\begin{pmatrix} 1 & 0 \\ a & 1 \end{pmatrix} \in E(r)$ and $E(R, \underline{A})$ is normal in $E(R)$ by its definition. But the proof of theorem 13.2 adapts to show that if $c \in GL(n, R, \underline{A})$, then $\begin{pmatrix} c & 0 \\ 0 & c^{-1} \end{pmatrix} \in E(2n, R, \underline{A})$. In particular, $\begin{pmatrix} b^{-1} & 0 \\ 0 & b \end{pmatrix} \in E(2n, R, \underline{A})$. But $\begin{pmatrix} ab & 0 \\ 0 & 1 \end{pmatrix}\begin{pmatrix} b^{-1} & 0 \\ 0 & b \end{pmatrix} = \begin{pmatrix} a & 0 \\ 0 & b \end{pmatrix}$.

Done with lemma 15.3.

Therefore $\begin{pmatrix} ab & 0 \\ 0 & 1 \end{pmatrix}^{-1} \begin{pmatrix} ba & 0 \\ 0 & 1 \end{pmatrix} \in E(2n, R, \underline{A})$. But $\begin{pmatrix} ab & 0 \\ 0 & 1 \end{pmatrix}^{-1} \begin{pmatrix} ba & 0 \\ 0 & 1 \end{pmatrix} = \begin{pmatrix} b^{-1}a^{-1}ba & 0 \\ 0 & 1 \end{pmatrix}$. Hence, $b^{-1}a^{-1}ba \in E(R, \underline{A})$ and $E(R, \underline{A}) \supset [GL(R), GL(R, \underline{A})]$.

Done with theorem 15.1 and corollary 15.2.

Therefore, all groups in the diagram at the beginning of this chapter are commutative. Hence, there is a well defined homomorphism $K_2(R/\underline{A}) \to K_1(R, \underline{A})$ such that the sequence

$$0 \to K_2(R, \underline{A}) \to K_2(R) \to K_2(R/\underline{A}) \to K_1(R, \underline{A}) \to K_1(R) \to K_1(R/\underline{A})$$

is exact. We summarize this in a theorem.

<u>Theorem 15.4</u>. Let R be a ring and $\underline{A}$ a 2 sided ideal of R. Then the sequence

$$0 \to K_2(R, \underline{A})' \to K_2(R) \to K_2(R/\underline{A}) \to K_1(R, \underline{A}) \to K_1(R) \to K_1(R/\underline{A})$$

is exact.

<u>Remark</u>. The difficulty with the definition of $K_2(R, \underline{A})'$ is that $K_2(R, \underline{A}) \to K_2(R)$ is a mono and so the sequence of K_i stops. We would like to have a long exact sequence involving K_n for all $n > 0$ (whatever these are). An alternative possibility is to define

$ST(R, \underline{A})$ to be generated by $y_{ij}(a)$ where $a \in \underline{A}$ and relations $y_{ij}(a + b) = y_{ij}(a)y_{ij}(b)$ for a and $b \in \underline{A}$ and

$$[y_{ij}(b), y_{k\ell}(b)] = \begin{cases} 1 & \text{if } i \neq \ell \text{ and } j \neq k \\ y_{i\ell}(b^2) & \text{if } i \neq \ell \text{ and } j = k. \end{cases}$$

and we let $ST(R)$ operate on $ST(R, \underline{A})$ by

$$x_{ij}(t)\bullet y_{k\ell}(a) = \begin{cases} y_{kl}(a) & \text{if } i \neq \ell \text{ and } j \neq k \\ y_{i\ell}(ta)y_{k\ell}(a) & \text{if } i \neq \ell \text{ and } j = k \\ y_{kj}(at)y_{k\ell}(a) & \text{if } i = \ell \text{ and } j \neq k. \end{cases}$$

Then we get homomorphisms $ST(R, \underline{A}) \longrightarrow GL(R, \underline{A})$ sending $y_{ij}(a)$ to $I + ae_{ij}$ and $ST(R, \underline{A}) \longrightarrow ST(R)$ sending $y_{ij}(a)$ to $x_{ij}(a)$. The latter has image $ST(R, A)'$. Call the kernel $K_2(R, \underline{A})$. For this to be an acceptable definition we would have to show it was abelian and define $K_3(R/\underline{A}) \longrightarrow K_2(R, \underline{A})$ to fit into the sequence but this has not yet been done.

Next we want to define maps and a relative K_0 so that the sequence $K_1(R/\underline{A}) \longrightarrow K_0(R, \underline{A}) \longrightarrow K_0(R) \longrightarrow K_0(R/\underline{A})$ is exact. We will define $K_0(R, f)$ where $f: R \longrightarrow R'$ is a ring homomorphism. No one has tried to define $K_1(R, f)$ yet.

Definitions. Let R and R' be rings and $f: R \longrightarrow R'$ a homomorphism. $\underline{F_f}$ is the category whose objects are triples (A, g, B) where A and B are finitely generated projective R modules and $g: R' \otimes_R A \longrightarrow R' \otimes_R B$ is an iso of R' modules. A map $(\ell, m): (A, g, B) \longrightarrow (A', g', B')$ is a pair of maps $\ell: A \longrightarrow A'$ and $m: B \longrightarrow B'$ such that the square

$$\begin{array}{ccc} R' \otimes_R A & \xrightarrow{\ell \otimes \ell} & R' \otimes_R A' \\ \downarrow & & \downarrow \\ R' \otimes_R B & \xrightarrow{\ell \otimes m} & R' \otimes_R B' \end{array} \quad \text{commutes.}$$

A sequence

$$0 \to (A', g', B') \to (A, g, B) \to (A'', g'', B'') \to 0$$

in F_f is <u>exact</u> if $0 \to A' \to A \to A'' \to 0$ and $0 \to B' \to B \to B'' \to 0$ are exact sequences of R modules.

$K_0(R.\ f)$ is the abelian group with generators the objects of F_f and relations

1) $[(A, g, B)] = [(A', g', B')] + [(A'', g'', B'')]$ if $0 \to (A', g', B') \to (A, g, B) \to (A'', g'', B'') \to 0$ is exact in F_f and

2) $[(A, gh, B)] = [(A, h, C)] + [(C, g, B)]$.

We map $K_0(R, f)$ to $K_0(R)$ by sending $[(A, g, B)]$ to $[A] - [B]$. We need to show this respects the relations. For relation of type 1) $0 \to A' \to A \to A'' \to 0$ and $0 \to B' \to B \to B'' \to 0$ exact implies $[A] - [B] = [A'] - [B'] + [A''] - [B'']$ in $K_0(R)$. Therefore relations of type 1) are preserved. For type 2) $[A] - [B] + [B] - [C] = [A] - [C]$ in $K_0(R)$ and hence relations of type 2) are preserved. Therefore, the map is well defined.

Now we want to define a map $K_1(R') \to K_0(R, f)$. We use the isomorphism of $K_1(R')$ with $GL(R')/E(R')$. If $g \in GL(n, R')$ send g to $[(R^n, g, R^n)]$ in $K_0(R, f)$. First we show this commutes with $GL(n, R') \to GL(n + 1, R')$. If $g \in GL(n, R')$, g goes to

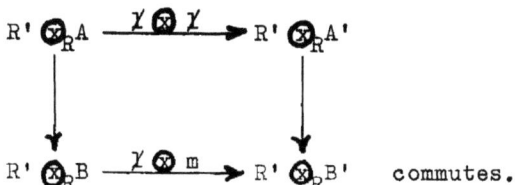

$\begin{pmatrix} g & 0 \\ 0 & 1 \end{pmatrix}$ in GL(n + 1, R'). $\begin{pmatrix} g & 0 \\ 0 & 1 \end{pmatrix}$ goes to $[(R^{n+1}, \begin{pmatrix} g & 0 \\ 0 & 1 \end{pmatrix}, R^{n+1})]$ in $K_0(R, f)$. But $[(R^{n+1}, \begin{pmatrix} g & 0 \\ 0 & 1 \end{pmatrix}, R^{n+1})] = [(R^n, g, R^n)] + [(R, 1, R)]$

$= [(R^n, g, R^n)]$ by the relation of type 2) (i.e., $[(A, 1, A)] = [(A, 1, A)] + [(A, 1, A)]$ using $1 \cdot 1 = 1$). This is a homomorphism of groups since gh goes to $[(R^n, gh, R^n)] = [(R^n, h, R^n)] + [(R^n, g, R^n)]$ by the relations of type 2). But $K_0(R, f)$ is an abelian group. Therefore, the map $GL(R') \longrightarrow K_0(R, f)$ factors through

$GL(R')/[GL(R'), GL(R')] = GL(R')/E(R') = K_1(R')$.

<u>Theorem 15.5</u>. Let $f: R \longrightarrow R'$ be a homomorphism of rings. Then the sequence

$$K_1(R) \longrightarrow K_1(R') \longrightarrow K_0(R, f) \longrightarrow K_0(R) \longrightarrow K_0(R')$$

constructed above is exact.

<u>Proof</u>. First we show the composition of any two consecutive maps is 0

1) Let $g \in GL(R)$. Then $g \rightsquigarrow f(g) \in GL(R')$ $f(g) \rightsquigarrow (R^n, f(g), R^n)$.

Then the diagram

$$\begin{array}{ccc} R' \otimes_R R^n & \xrightarrow{1 \otimes g} & R' \otimes_R R^n \\ \downarrow{\scriptstyle f(g)} & & \downarrow{\scriptstyle 1} \\ R' \otimes_R R^n & \xrightarrow{1 \otimes 1} & R' \otimes_R R^n \end{array}$$

is a commutative square of isomorphisms. Therefore, $(R^n, f(g), R^n)$ is isomorphic to $(R^n, 1, R^n)$ in F_f. Therefore

$[(R^n, f(g), R^n)] = [(R^n, 1, R^n)] = 0$ in $K_0(R, f)$.

2) The composition $K_1(R') \longrightarrow K_0(R, f) \longrightarrow K_0(R)$ sends $g \in GL(n, R')$ to $[R^n] - [R^n] = 0$ in $K_0(R)$.

3) $(A, g, B) \in K_0(R, f)$ goes to $[R' \otimes_R A] - [R' \otimes_R B]$ in $K_0(R')$. But $g: R' \otimes_R A \longrightarrow R' \otimes_R B$ is an iso. Therefore, $[R' \otimes_R A] - [R' \otimes_R B] = 0$ in $K_0(R')$.

Now we show that the kernels are contained in the images.

4) Let $x \in K_0(R)$ go to 0 in $K_0(R')$. $x = [P] - [Q]$ has image $[P \otimes_R R'] - [Q \otimes_R R'] = 0$. Therefore, there exists an n such that $(R' \otimes_R P) \oplus R'^n$ is isomorphic to $(Q \otimes_R R') \oplus R'^n$ and $x = [P \oplus R^n] - [Q \oplus R^n]$. That is, we can choose P and Q such that there is an iso $g: R' \otimes_R P \longrightarrow R' \otimes_R Q$ and then $(P, g, Q) \in K_0(R, f)$ has image x in $K_0(R)$.

5) Let $x \in K_0(R, f)$ go to 0 in $K_0(R)$. Then we want that x is in the image of $K_1(R')$. First we need a lemma.

Lemma 15.6. Every element of $K_0(R, f)$ has the form $[(A, g, B)]$.

Proof. $[(A, g, B)] + [(A', g', B')] = [(A \oplus A', g \oplus g', B \oplus B')]$. Also $-[(A, g, B)] = [(B, g^{-1}, A)]$ by the relations of type 2. Therefore, sums of the generators and negatives of the generators can be put in the given form. Hence every element can be put in that form. Done with lemma 15.6.

Therefore $x = [(A, g, B)]$ and $0 = [A] - [B]$ in $K_0(R)$. Therefore there is an n such that $A \oplus R^n$ is isomorphic to $B \oplus R^n$. But then $x = [(A \oplus R^n, g \oplus 1, B \oplus R^n)]$. By changing notation we can assume that B is isomorphic to A by an iso h. But then the pair of maps $1: A \longrightarrow A$ and $h: B \longrightarrow A$ show that $(A, g, B) \cong (A, (R' \otimes h)g, A)$ in $K_0(R, f)$. Hence we can assume $A = B$. We will formulate the rest of the argument as a lemma since it will be used again later.

Lemma 15.7. If P is a finitely generated projective R-module and g: $R' \otimes_R P \longrightarrow R' \otimes_R P$ is an isomorphism, then $[(R' \otimes_R P, g)] \in K_1(R')$ maps onto $[(P, f, P)] \in K_0(R, f)$.

Proof. Let $P \oplus Q = F = R^n$. Then $[(R' \otimes_R Q, 1)] = 0$ so $[(R' \otimes_R F, g \oplus 1)] = [(R' \otimes_R P, g)]$. Similarly, $[(Q, 1, Q)] = 0$. Thus, $[(F, g \oplus 1, F)] = [(P, g, P)]$. Therefore, we can assume $P = R^n$ in which case the result is just the definition of $K_1(R') \longrightarrow K_0(R, f)$. Done with lemma 15.7.

6) Let $x \in K_1(R')$ which goes to 0 in $K_0(R, f)$. Then we want to show that x is in the image of $K_1(R)$. First we need to simplify the relations we have in $K_0(R, f)$. By lemma 15.6 every element of $K_0(R, f)$ has the form $[(A, g, B)]$.

Lemma 15.8. $[(A, g, B)] = [(A', g', B')]$ in $K_0(R, f)$ if and only if there exists C and C', finitely generated projective R modules, and $h: (B \oplus C) \otimes_R R' \longrightarrow (B \oplus C) \otimes_R R'$, an R' iso, such that $(A \oplus C, h(g \oplus 1), B \oplus C)$ is isomorphic to $(A' \oplus C', g' \oplus 1, B' \oplus C')$ in F_f and such that $[((B \oplus C) \otimes_R R', h)] = 0$ in $K_1(R')$.

Proof. $\Longleftarrow$ $[(A \oplus C, h(g \oplus 1), B \oplus C)] = [(A \oplus C, (g \oplus 1), B \oplus C)] + [(B \oplus C, h, B \oplus C)] = [(A, g, B)] + [(B \oplus C, h, B \oplus C)]$. $[(A' \oplus C', g' \oplus 1, B' \oplus C')] = [(A', g', B')]$ in $K_0(R, f)$. Therefore, it is enough to show that $[(B \oplus C, h, B \oplus C)] = 0$ in $K_0(R, f)$. There exists Q such that $B \oplus C \oplus Q \cong R^n$ and $[(B \oplus C \oplus Q, h \oplus 1, B \oplus C \oplus Q)] = [(R^n, \chi, R^n)] = [(B \oplus C, h, B \oplus C)]$. In $K_1(R')$ we have $[(R^n \otimes_R R', \chi)] = [(B \oplus C \oplus Q) \otimes_R R', h \oplus 1)] = [((B \oplus C) \otimes_R R', h)] + [(Q \otimes_R R', 1)] = 0 + 0 = 0$. The first by hypothesis, and the second is clear. But $[(R^n \otimes_R R', \chi)]$ goes to $[(B \oplus C, h, B \oplus C)]$ in $K_0(R, f)$. Therefore, $[(B \oplus C, h, B \oplus C)] = 0$ as desired, and $[(A, g, B)] = [(A', g', B')]$.

$\Longrightarrow$ We take the relation on $[(A, g, B)]$ and $[(A', g', B')]$ defined by the lemma. We show that it is an equivalence relation, that we can make a group out of the equivalence classes, and that this group is $K_0(R, f)$.

1) The relation is an equivalence relation.

a) It is clearly reflexive.

b) Suppose that $(A \oplus C, h(g \oplus 1), B \oplus C)$ is isomorphic to $(A' \oplus C', g' \oplus 1, B' \oplus C')$. Hence we have isos s and r such that the diagram

$$\begin{array}{ccc}
(A \oplus C) \otimes_R R' & \xrightarrow{\ r\ } & (A' \oplus C') \otimes_R R' \\
\downarrow{\scriptstyle g \oplus 1} & & \downarrow{\scriptstyle g' \oplus 1} \\
(B \oplus C) \otimes_R R' & & (B' \oplus C') \otimes_R R' \\
\downarrow{\scriptstyle h} & & \downarrow{\scriptstyle 1} \\
(B \oplus C) \otimes_R R' & \xrightarrow{\ s\ } & (B' \oplus C') \otimes_R R'
\end{array}$$

commutes.

We want to define h' such that $[((B' \oplus C') \otimes_R R', h')] = 0$ in $K_1(R')$ and such that

$$\begin{array}{ccc}
(A' \oplus C') \otimes_R R' & \xrightarrow{\ r^{-1}\ } & (A \oplus C) \otimes_R R' \\
\downarrow{\scriptstyle g' \oplus 1} & & \downarrow{\scriptstyle g \oplus 1} \\
(B' \oplus C') \otimes_R R' & & (B \oplus C) \otimes_R R' \\
\downarrow{\scriptstyle h'} & & \downarrow{\scriptstyle 1} \\
(B' \oplus C') \otimes_R R' & \xrightarrow{\ s^{-1}\ } & (B \oplus C) \otimes_R R'
\end{array}$$

commutes.

That is, we know that $1(g' \oplus 1)r = sh(g \oplus 1)$. We want

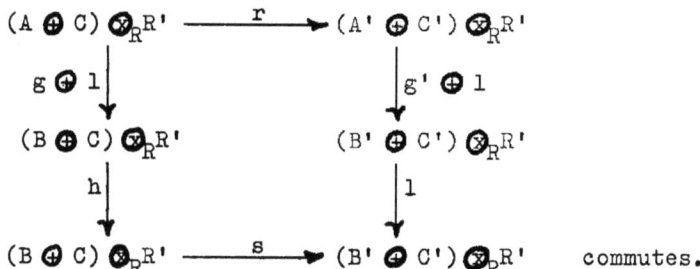

$s^{-1}h'(g' \oplus 1) = 1(g \oplus 1)r^{-1}$. Equivalently $h'(g' \oplus 1)r = s1(g \oplus 1)$ since s and r are isos. But $h'(g' \oplus 1)r = h'sh(g \oplus 1)$. Hence, it is enough that $h'sh = s$ or that $h' = sh^{-1}s^{-1}$. Define h' by $h' = sh^{-1}s^{-1}$. Then the diagram commutes and h' goes to 0 in $K_1(R')$ since h does. Therefore, the relation is symmetric.

c) Finally we show the relation is transitive.

We have the commutativity of the diagrams

$$\begin{array}{ccc}
(A \oplus C) \otimes_R R' & \xrightarrow{r} & (A' \oplus C') \otimes_R R' \\
\downarrow{\scriptstyle g \oplus 1} & & \downarrow{\scriptstyle g' \oplus 1} \\
(B \oplus C) \otimes_R R' & & (B' \oplus C') \otimes_R R' \\
\downarrow{\scriptstyle h} & & \downarrow{\scriptstyle 1} \\
(B \oplus C) \otimes_R R' & \xrightarrow{s} & (B' \oplus C') \otimes_R R'
\end{array}
\qquad
\begin{array}{ccc}
(A' \oplus C'') \otimes_R R' & \xrightarrow{r'} & (A'' \oplus C''') \otimes_R R' \\
\downarrow{\scriptstyle g' \oplus 1} & & \downarrow{\scriptstyle g'' \oplus 1} \\
(B' \oplus C'') \otimes_R R' & & (B'' \oplus C''') \otimes_R R' \\
\downarrow{\scriptstyle h'} & & \downarrow{\scriptstyle 1} \\
(B' \oplus C'') \otimes_R R' & \xrightarrow{s'} & (B'' \oplus C''') \otimes_R R'.
\end{array}$$

By adding C" to the left hand diagram and C' to the right hand diagram we can assume that C' and C" are the same. Then the map $h'' = s^{-1}h'sh: (B \oplus C) \otimes_R R' \to (B \oplus C) \otimes_R R'$ makes the diagram

$$\begin{array}{ccc}
(A \oplus C) \otimes_R R' & \xrightarrow{r'r} & (A'' \oplus C''') \otimes_R R' \\
\downarrow{\scriptstyle g \oplus 1} & & \downarrow{\scriptstyle g'' \oplus 1} \\
(B \oplus C) \otimes_R R' & & (B'' \oplus C''') \otimes_R R' \\
\downarrow{\scriptstyle h''} & & \downarrow{\scriptstyle 1} \\
(B \oplus C) \otimes_R R' & \xrightarrow{s's} & (B'' \oplus C''') \otimes_R R'
\end{array}$$

commutes.

h" goes to 0 in $K_1(R')$ since $s^{-1}h's$ does and h does and products of things going to 0 go to 0. Therefore, the relation is transitive, and the relation is an equivalence relation.

2) Let G be the set of equivalence classes. Then G is a commutative monoid with respect to $\oplus$. It is enough to show that if a triple T is equivalent to T' and S is a triple then $S \oplus T$ is equivalent to $S \oplus T'$. This is clear.

3) G is an abelian group. [(A, 1, A)] is certainly the identity. Therefore, all we need is inverses. We claim $[(B, -g^{-1}, A)]$ is the inverse of [(A, g, B)]. $(A, g, B) \oplus (B, -g^{-1}, A) = (A \oplus B, g \oplus -g^{-1}, B \oplus A) \cong (A \oplus B, \begin{pmatrix} 0 & g \\ -g^{-1} & 0 \end{pmatrix}, A \oplus B)$. Hence by lemma 15.7 it is enough to show that $\begin{pmatrix} 0 & g \\ -g^{-1} & 0 \end{pmatrix}$ represents 0 in $K_1(R')$. But $\begin{pmatrix} 0 & g \\ -g^{-1} & 0 \end{pmatrix} = \begin{pmatrix} 1 & 0 \\ -g^{-1} & 1 \end{pmatrix}\begin{pmatrix} 1 & g \\ 0 & 1 \end{pmatrix}\begin{pmatrix} 1 & 0 \\ -g^{-1} & 1 \end{pmatrix}$. Therefore it is enough to show matrices of the form $\begin{pmatrix} 1 & x \\ 0 & 1 \end{pmatrix}$ and $\begin{pmatrix} 1 & 0 \\ y & 1 \end{pmatrix}$ represent 0. But the exact sequence

$$0 \longrightarrow (Q, 1) \longrightarrow (P \oplus Q, \begin{pmatrix} 1 & x \\ 0 & 1 \end{pmatrix}) \longrightarrow (P, 1) \longrightarrow 0$$

shows $\begin{pmatrix} 1 & x \\ 0 & 1 \end{pmatrix}$ represents 0 in $K_1(R')$. $\begin{pmatrix} 1 & 0 \\ y & 1 \end{pmatrix}$ follows by symmetry

Hence $\begin{pmatrix} 0 & g \\ -g^{-1} & 0 \end{pmatrix}$ represents 0 and $[(B, -g^{-1}, A)]$ is the inverse of [(A, g, B)].

<u>Remark</u>. It is clear that (A, g, B) is equivalent to (A, -g, B).

4) Next we check that the group G satisfies the relation for triples in $K_0(R, f)$. Let $0 \longrightarrow S' \longrightarrow S \longrightarrow S'' \longrightarrow 0$ be an exact sequence of triples. Let (P', g', Q'). (P, g, Q), and (P", g", Q") be representatives. Then $0 \longrightarrow P' \longrightarrow P \longrightarrow P'' \longrightarrow 0$ exact and

$0 \to Q' \to Q \to Q'' \to 0$ exact implies that $P \cong P' \oplus P''$ and $Q \cong Q' \oplus Q''$. Hence we can assume $S = (P' \oplus P'', \begin{pmatrix} g' & h \\ 0 & g'' \end{pmatrix}, Q' \oplus Q'')$. $S' \oplus S'' = (P' \oplus P'', \begin{pmatrix} g' & 0 \\ 0 & g'' \end{pmatrix}, Q' \oplus Q'')$. But $\begin{pmatrix} g' & h \\ 0 & g'' \end{pmatrix} = \begin{pmatrix} g' & 0 \\ 0 & g'' \end{pmatrix}\begin{pmatrix} 1 & g'^{-1} \\ 0 & 1 \end{pmatrix}$ and $[P \oplus P'', \begin{pmatrix} 1 & g'^{-1}h \\ 0 & 1 \end{pmatrix}, P \oplus P''] = 0$ as above (using lemma 15.7). Therefore, $S' \oplus S''$ is equivalent to S. Hence G satisfies the relations on exact sequences.

5) Finally we check that $SS' = S \oplus S'$ when SS' is defined. Let $S = (P, g, Q)$ and $S' = (Q, h, M)$. Then S' is equivalent to $(Q, -h, M)$. We want $(P, hg, M) \oplus (Q, 1, Q)$ equivalent to $(P \oplus Q, \begin{pmatrix} g & 0 \\ 0 & h \end{pmatrix}, Q \oplus M)$. But $\begin{pmatrix} 0 & 1 \\ 1 & 0 \end{pmatrix}\begin{pmatrix} g & 0 \\ 0 & h \end{pmatrix} = \begin{pmatrix} 0 & -h \\ g & 0 \end{pmatrix} = \begin{pmatrix} hg & 0 \\ 0 & 1 \end{pmatrix}\begin{pmatrix} 1 & -g^{-1} \\ 0 & 1 \end{pmatrix}\begin{pmatrix} 1 & 0 \\ g & 1 \end{pmatrix}$ and $\begin{pmatrix} 0 & -1 \\ 1 & 0 \end{pmatrix} \in E(R)$.

Therefore SS' is equivalent to $S \oplus S'$. We use lemma 15.7 as above to eliminate the various elementary matrices.

6) G is an abelian group which satisfies all the relations and $G \to K_0(R, f)$ given by $S \to S$ where S is any triple is a well defined homomorphism of groups. By the universality of $K_0(R, f)$ there is a unique map $K_0(R, f) \to G$. It is clear this is an iso, since the map $G \to K_0(R, f)$ by $(A, f, B) \to [(A, f, B)]$ is a homomorphism inverse to $K_0(R, f) \to G$. Done with lemma 15.8.

We can now easily prove theorem 15.5. Let $x \in K_1(R')$ go to 0 in $K_0(R, f)$. Let (R^n, g) represent x. Then $[(R^n, g, R^n)] = 0$ in $K_0(R, f)$. Apply lemma 15.8 to $(A, g, B) = (0, 0, 0)$, $(A', g', B') = (R^n, g, R^n)$. This gives us C, C', $h: R' \otimes_R C \cong R' \otimes_R C$ so $(C, h, C) \cong (R^n \oplus C', g \oplus 1, R^n \oplus C')$ and such that $[(R' \otimes_R C, h)] = 0$ in $K_1(R')$. Let $C' \oplus C'' \cong R^m$. Then

$(C \oplus C'', h \oplus 1, C \oplus C'') \cong (R^n \oplus R^m, g \oplus 1, R^n \oplus R^m)$ by adding $(C'', 1, C'')$ to both sides. Now $g \oplus 1$ still represents x in $K_1(R')$ and $h \oplus 1$ represents the same element as h, namely 0. Changing notation, write g for $g \oplus 1$, h for $h \oplus 1$, R^n for $R^n \oplus R^m$, C for $C \oplus C''$. Thus $(R^n, g, R^n) \cong (C, h, C) \cong (R^n, h, R^n)$ since the isomorphism implies $C \cong R^n$. Here g represents x and h represents 0 in $K_1(R)$. The isomorphism is given by two isomorphisms $k: R^n \cong R^n$, $\chi: R^n \cong R^n$ such that the diagram

$$\begin{array}{ccc} R' \otimes R^n & \xrightarrow{R' \otimes k} & R' \otimes R^n \\ \downarrow g & & \downarrow h \\ R' \otimes R^n & \xrightarrow{R' \otimes \chi} & R' \otimes R^n \end{array}$$

commutes. Therefore $g = (R' \otimes \chi)^{-1} h (R' \otimes k)$. In $K_1(R')$, h represents 0 and $R' \otimes \chi$, $R' \otimes k$ represent the images of $[(R^n, \chi)]$, $[(R^n, k)] \in K_1(R)$. Thus x is the image of $[(R^n, k)] - [(R^n, \chi)] \in K_1(R)$. Done with theorem 15.5.

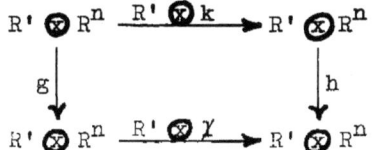

Chapter 16. Further Results on K_1 and K_0.

In this chapter we present additional results on K_1 and K_0. In particular we discuss $K_1(R[X])$ and, $K_1(R[X, X^{-1}])$.

<u>Definition</u>. Let R be a ring. Then $A \in GL(R)$ is <u>unipotent</u> if $A - I$ is a nilpotent matrix.

<u>Remark</u>. If N is an $n \times n$ nilpotent matrix, then $I + N$ is invertible with inverse $I - N + N^2 - N^3 + \ldots$.

Let R be a ring and X an indeterminate. Then $R \longrightarrow R[X]$ by the inclusion and $R[X] \longrightarrow R$ by $X \longrightarrow 0$ give map $K_1(R[X]) \rightleftarrows K_1(R)$. This splits $K_1(R)$ off as a summand of $K_1(R[X])$.

<u>Theorem 16.1</u>. Let R be a ring and X an indeterminate. Then $K_1(R[X]) \cong K_1(R) \oplus V$ where V is the subgroup of $K_1(R[X])$ generated by all $I + xN$ with N a nilpotent matrix with all its entries in R.

<u>Theorem 16.2</u>. If R is left regular and $A \in GL(R)$ is unipotent, then A goes to 0 in $K_1(R)$.

<u>Corollary 16.3</u>. If R is left regular, then $K_1(R)$ is isomorphic to $K_1(R[X])$.

<u>Proof of theorem 16.1</u>. We have $V \subset K_1(R[X])$ and $K_1(R) \subset K_1(R[X])$. First we show this sum is direct. Let

$$0 \longrightarrow V' \longrightarrow K_1(R[X]) \leftrightarrows K_1(R) \longrightarrow 0 \text{ be exact.}$$

Then $I + xN \in V$ goes to I in $K_1(R)$ since x goes to 0 so $V \subset V'$. Hence the sum is direct and $V = V'$ if and only if $V + K_1(R) = K_1(R[X])$. Let $A \in GL(R[X])$. Then A is represented by $A' \in GL(n, R[X])$ and $A' = A_0 + A_1x + \ldots + A_mx^m$ where the A_i have all their entries in R. We claim that there exists an $E \in E(R[X])$ such that

$AE = \begin{pmatrix} B_0+B_1x & 0 \\ 0 & I \end{pmatrix}$ where B_0 and B_1 have all their entries in R. We prove this by induction on m. If m = 1 we are done. Let I be an $n \times n$ identity then

$$\begin{pmatrix} A' & 0 \\ 0 & I \end{pmatrix} \begin{pmatrix} I & X \\ 0 & I \end{pmatrix} = \begin{pmatrix} A' & A'X \\ 0 & I \end{pmatrix} .$$

Since A' is invertible A'X can be any $n \times n$ matrix. Let $A'X = A_m X^{m-1}$. Then

$$\begin{pmatrix} A' & A_m X^{m-1} \\ 0 & I \end{pmatrix} \begin{pmatrix} I & 0 \\ Y & I \end{pmatrix} = \begin{pmatrix} A + A_m X^{m-1} Y & A_m X^{m-1} \\ Y & I \end{pmatrix} .$$

Let Y = -xI. Then the result is

$$\begin{pmatrix} A_0+A_1x + \dots + A_{m-1}x^{m-1} & A_m x^{m-1} \\ -xI & I \end{pmatrix}$$

and all entries have lower powers of X. Hence we are finished with the claim. We cannot remove all powers of X by this method since we introduce an X term in the bottom. Now $B_0 + B_1x \in GL(R[X])$ goes to $B_0 \in GL(R)$. Hence $B_0 \in GL(R[X])$ and is an element in the image of $K_1(R) \longrightarrow K_1(R[X])$. Let $C = B^{-1}B_1$. Then $K_1(R[X])$ is generated by $K_1(R)$ and all $I + xC \in GL(R[X])$. If we can show that C is nilpotent, we will be finished with theorem 16.1. I + xC is invertible. Therefore, there exist D_i whose entries are in R such that

$(I + xC)(D_0 + D_1x + \dots + D_rx^r) = I.$ Therefore

$D_0 = I$

$CD_0 + D_1 = 0$

$CD_1 + D_2 = 0$

- - - - -

$CD_{r-1} + D_r = 0$

$CD_r = 0$ by comparing both sides.

Therefore, $D_1 = -CD_0$

$D_2 = -CD_1$

- - - - -

$D_r = -CD_{r-1}$.

But this implies

$D_r = (-C)^r D_0$ and $D_r = (-C)^r$. Then

$CD_r = 0 = (-1)^r C^{r+1}$ and C is nilpotent.

Done with theorem 16.1.

Remark. Note that we have shown that every element of V is represented by a matrix $I + xN$ with N nilpotent over R.

We generalize this to $K_1(R[X, X^{-1}])$. We map $R \longrightarrow R[X, X^{-1}]$ by inclusion and $R[X, X^{-1}] \longrightarrow R$ by sending $X \rightsquigarrow 1$. Then this splits $K_1(R)$ off as a summand of $K_1(R[X, X^{-1}])$.

Theorem 16.4. Let R be any ring. Then $K_1(R[X, X^{-1}])$ is isomorphic to $K_1(R) \oplus K_0(R) \oplus W$ where W is generated by all $I + (X-1)N$ and $I + (X^{-1}-1)N$ where N is nilpotent with entries in R.

Corollary 16.5. If R is left regular, $K_1(R[X, X^{-1}])$ is isomorphic to $K_1(R) \oplus K_0(R)$.

Proof. Immediate from theorem 16.2 and theorem 16.4.

Notation. If P is an R module, let $P[X] = R[X] \otimes_R P$.

Proof of theorem 16.4. First we need to show how $K_0(R)$ is imbedded in $K_1(R[X, X^{-1}])$. We map $K_0(R) \to K_1(R[X, X^{-1}])$ by sending [P] to $[(P[X, X^{-1}]$, multiplication by X)]. This clearly preserves relations. We now define a left inverse of this map. Let $A \in GL(R[X, X^{-1}])$. Let $A' \in GL(n_1 R[X, X^{-1}])$ be a representative. If all entries of A' lie in R[X] then we have

$R[X]^n \subset R[X, X^{-1}]^n$ and A' induces

$A'^*: R[X]^n \to R[X]^n$. A'^* is a mono since A' is. We form the exact sequence

$$0 \to R[X]^n \xrightarrow{A'^*} R[X]^n \to P'(A) \to 0.$$ It

is clear that if $A'' \in GL(m, R[X, X^{-1}])$ was another representative of A then P'(A) would be the same.

Lemma 16.6. If A and $B \in GL(R[X, X^{-1}])$ and A and B have all their entries in R[X], then the sequence

$0 \to P'(B) \to P'(AB) \to P'(A) \to 0$ is exact.

Proof. Pick n big enough so that A and B are represented in $GL(n, R[X, X^{-1}])$ by A' and B' respectively. Let $F = R[X]^n$. Then we have

$$\begin{array}{ccccccccc} & & & & P(AB) & & P(A) & & \\ & & & & \| & & \| & & \\ 0 & \to & A'^*F/(A'B')^*F & \to & F/(A'B')^*F & \to & F/A'^*F & \to & 0 \end{array}$$ where

the rows is exact. Hence it is enough to show that $A'^*F/(A'B')^*F$ is isomorphic to $P'(B) = F/B'^*F$. A'^* is a mono since A' is. We have the commutative diagram

$$\begin{array}{ccc} A'^*F & \xleftarrow{\ i\ } & (A'B')^*F \\ {\scriptstyle A'^*}\uparrow\cong & & \cong\uparrow{\scriptstyle (A'B')^*} \\ F & \xleftarrow[B'^*]{} & F \end{array}$$

where i is the inclusion.

Therefore, the cokers $F \xrightarrow{B'^*} F$ and $(A'B')^*: F \longrightarrow A'^*F$ are isomorphic. That is $A'^*F/(A'B')^*F \cong F/B'^*F$.

Done with lemma 16.6.

<u>Lemma 16.7</u>. If $A \in GL(R[X, X^{-1}])$ and has all its entries in $R[X]$, then $P'(A)$ is a finitely generated projective R module.

<u>Proof</u>. $R[X, X^{-1}]$ is a free R module with generators X^n where $n \in Z$. $R[X]$ is a free R module with generators X^n where $n = 0, 1, 2, \ldots$. Define $j: R[X, X^{-1}] \longrightarrow R[X]$ by $j(X^n) = X^n$ if $n \geq 0$ and $j(X^n) = 0$ if $n < 0$. Let $i: R[X] \longrightarrow R[X, X^{-1}]$ be the inclusion. Then $ji = 1_{R[X]}$. Let A be represented by $A' \in GL(n, R[X, X^{-1}])$. Then $(j \oplus \ldots \oplus j)A'^{-1}(i \oplus \ldots \oplus i): R[X]^n \longrightarrow R[X]^n$ is a map of R modules splitting the monomorphism $A'^*: R[X]^n \longrightarrow R[X]^n$. Therefore, coker $(A'^*) = P'(A)$, is a projective R module.

To show $P'(A)$ is finitely generated it is enough to show that $X^m P'(A) = 0$ for some m. Then $P'(A)$ will be a quotient of $R[X]^n/X^m R[X]^n$ which is finitely generated over R. Now $A'^{-1} \in GL(n, R[X, X^{-1}])$ so $B = X^m A$ has entries in $R[X]$ for some m. Therefore, setting $F = R[X]^n$, we have $A'^*F \supset A'^*BF = X^m F$.

Done with lemma 16.7.

<u>Definition</u>. If $A \in GL(R[X, X^{-1}])$, then there exist

$$T = \begin{pmatrix} X^{n_1} & & & & & 0 \\ & \ddots & & & & \\ & & X^{n_m} & & & \\ & & & 1 & & \\ 0 & & & & \ddots & \\ & & & & & 1 \end{pmatrix} \in GL(R[X, X^{-1}])$$ such that TA has all its entries in R[X]. Let <u>P(A)</u> $= [P'(TA)] - [P'(T)] \in K_0(R)$.

<u>Lemma 16.8</u>. 1) P(A) depends only on A (and hence P(A) = P'(A) when P'(A) is defined since P'(I) = 0).

2) $P: GL(R[X, X^{-1}]) \longrightarrow K_0(R)$ is a homomorphism of groups.

<u>Proof</u>. 1) Let $A \in GL(R[X, X^{-1}])$ and let T and T' be as above. We can assume A, T, and $T' \in GL(n, R[X, X^{-1}])$. Pick N big enough so that all powers of X in T and T' are less than N. Let I be the $n \times n$ identity. Then there exist T" and T‴ as above such that $TT'' = T'T''' = X^N I$. By lemma 16.6

$$[P'(TT''A)] = [P'(T''(TA))] = [P'(T'')] + [P'(TA)] \text{ and}$$
$$[P'(TT'')] = [P'(T'')] + [P'(T)].$$

Subtracting shows that TT" and T yield the same P(A). Similarly $T'T''' = TT''$ yields the same P(A) as T'. Therefore, T and T' yield the same and P is well defined.

2) Let A and $B \in GL(R[X, X^{-1}])$. Pick n big enough so that A and $B \in GL(n, R[X, X^{-1}])$. Pick T and S as above where TA has all its entries in R[X] and SB has all its entries in R[X]. Then TASB = TSAB. Therefore, P(AB) = [P(TSAB)] - [P(TS)] = [P(TASB)] - [P(TS)] = [P(TA)] + [P(SB)] - [P(T)] - [P(S)] = P(A) + P(B).

Done with lemma 16.8.

Now $K_0(R)$ is an abelian group. Therefore, P factors through $GL(R[X, X^{-1}])/E(R[X, X^{-1}]) = K_1(R[X, X^{-1}])$.

Lemma 16.9. Using the maps constructed above the composition are as follows:

a) $K_1(R) \to K_1(R[X, X^{-1}]) \to K_1(R)$ is 1,

b) $K_1(R) \to K_1(R[X, X^{-1}]) \to K_0(R)$ is 0,

c) $K_0(R) \to K_1(R[X, X^{-1}]) \to K_0(R)$ is 1,

d) $K_0(R) \to K_1(R[X, X^{-1}]) \to K_1(R)$ is 0,

e) $w \to K_1(R[X, X^{-1}]) \to K_1(R)$ is 0, and

f) $w \to K_1(R[X, X^{-1}]) \to K_0(R)$ is 0.

Remark. Given this and that $K_0(R)$, $K_1(R)$, and W generate $K_1(R[X, X^{-1}])$, the proof of theorem 16.4 is immediate.

Proof. a) is already done.

b) $A \in GL(R)$ goes to $A \in GL(R[X, X^{-1}])$. A already has its entries in R and is an R iso. Therefore, A'^* is an R iso and the coker $(A'^*) = 0$. Hence, the composite is 0.

c) $[P] \in K_0(R)$ goes to $[(Px, x^{-1}], x)]$. There exists Q such that $P \oplus Q$ is a finitely generated free R module. $[(P[x, x^{-1}], x)] = [(P[x, x^{-1}] \oplus Q[x, x^{-1}], x \oplus 1)]$. Let $x \oplus 1$ correspond to A on $F[x, x^{-1}]$ where $F = P \oplus Q$ is free with a chosen base. Then A has all its entries in R[X]. Hence it is enough to know the coker of $P[x] \oplus Q[x] \xrightarrow{x \oplus 1} P[x] \oplus Q[x]$. The coker of 1 is clearly 0. The coker of x is isomorphic to P since $P_0 + P_1x + \dots + P_nx^n \in P[x]$ where $P_i \in P$ is congruent mod x to P_0

and $P_0 \in P$ cannot be written px with $p \in P$. Hence [P] goes to [P].

d) [P] goes to $[(P[x, x^{-1}], x)]$. But x goes to 1 in GL(R). Therefore, $[(P[x, x^{-1}], x)]$ goes to $[(P, 1)] = 0$.

e) Since x goes to 1, anything of the form $I + (x-1)N$ or $I + (x^{-1}-1)N$ goes to $I \in GL(R)$. Therefore W goes to 0 in $K_1(R)$.

f) $I + (x-1)N$ is an R[X] isomorphism since N is nilpotent. Therefore, the coker of $I + (x-1)N$ is 0. Let $I + (x^{-1}-1)N \in GL(n, R[X, X^{-1}])$ where N is a nilpotent matrix in R. Then $xI(I + (x^{-1}-1)N) = xI + (1-x)N$ has all its entries in R[X]. $P(I + (x^{-1}-1)N) = P(xI + (1-x)N) - P(xI)$. $P(xI) = R^n$ is as in part c). Write $xI + (1-x)N = x(I - N) + N$. Since $I - N$ is an R iso, we can replace $x(I - N) + N$ by $xI + N(I - N)^{-1}$. Hence it is enough to show that if Q is any R module and $f: Q \to Q$ is any endomorphism, then $A[X] \xrightarrow{xI + f} Q[X] \xrightarrow{p} Q \to 0$ is an exact sequence of R modules where $p(\sum q_i x^i) = \sum(-1)^i f^i(q_i)$. This is clear since if $q_0 + \dots + q_n x^n \in Q[x]$ with the $q_i \in Q$ then

$$q_0 + \dots + q_n x^n \equiv \sum_{i=0}^{n} (-1)^i f^i(q_i) \text{ mod the image of } xI + f. \text{ Thus}$$

$P(xI + (x^{-1} - 1)N) = [R^n] - [R^n] = 0$. Done with lemma 16.9.

<u>Lemma 16.10</u>. $K_1(R)$, $K_0(R)$, and w generate $K_1(R[X, X^{-1}])$.

<u>Proof</u>. Let $A \in GL(n, R[X, X^{-1}])$ and let I be the $n \times n$ identity. Then there is an N big enough so that $(x^N I)(A)$ has all entries in R[X]. First we claim that $x^N I$ is in the image of $K_0(R)$. $x^N I$ is a product of matrices of the form $\begin{pmatrix} 1 & & & 0 \\ & \ddots & & \\ & & x^N & \\ 0 & & & \ddots \; 1 \end{pmatrix}$. But matrices of the

form $\begin{pmatrix} x & 0 \\ 0 & x^{-1} \end{pmatrix}$ are in $E(R[X, X^{-1}])$. Hence $x^N I \equiv \begin{pmatrix} x_1 & & \\ & \ddots & \\ & & 1 \end{pmatrix}^{N_n}$ mod $E(R[X, X^{-1}])$. But $[R] \in K_0$ goes to $[(R[X, X^{-1}], X)]$ in $K_1(R[X, X^{-1}])$ which is represented by the matrix $\begin{pmatrix} x_1 & & \\ & \ddots & \\ & & 1 \end{pmatrix}$. Therefore, $x^N I$ is in the image of $K_0(R)$ and so we can assume that A has all its entries in $R[X]$.

Next we claim there exist E and $E' \in E(R[X])$ such that

$$EAE' = \begin{pmatrix} A_0+A_1x & & 0 \\ & 1 & \\ 0 & & \ddots_1 \end{pmatrix}$$ in $GL(R[X, X^{-1}])$ where A_0 and A_1 have entries in R. Let $A = B_0 + \ldots + B_nX^n$ where B_i have entries in R, then

$$\begin{pmatrix} I & -xI \\ 0 & I \end{pmatrix}\begin{pmatrix} A & 0 \\ 0 & I \end{pmatrix}\begin{pmatrix} I & 0 \\ B_nX^{n-1} & 1 \end{pmatrix} = \begin{pmatrix} A-B_nX^n & -xI \\ B_nX^{n-1} & I \end{pmatrix},$$ and

this matrix has no powers of x greater than $n - 1$ occuring. The claim is finished by induction.

Therefore, we can assume $A = A_0 + A_1x = A_0 + A_1 + A_1(x-1)$ where A_i have all their entries in R. Since $x \rightsquigarrow 1$ in the map $GL(R[X, X^{-1}]$ to $GL(R)$, we know that $A_0 + A_1 \in GL(R)$. If $B \in GL(R)$, then we can use $B^{-1}A$ in place of A.

Let $B = A_0 + A_1$. Then $B^{-1}A$. $C = I + (x-1)B^{-1}A$. Let $C = B^{-1}A_1$. Then it is enough to prove the lemma in the case $A = I + (x-1)C$.

Let $D_{-q}x^q + \ldots + D_0 + D_1x + \ldots + D_px^p = A^{-1}$ where all D_i have entries in R. Then $I + (x-1)C = (I - C) + xC$. Therefore,

$\sum D_i x^i(I - C) + \sum D_{i-1}x^i C = I$. Comparing the coefficients of x we have:

$D_i(I - C) + D_{i-1}C = \begin{matrix} 0 \text{ if } i \neq 0 \\ I \text{ if } i = 0. \end{matrix}$ That is

$D_{-q}(I - C) = 0$ $\qquad$ $D_0(I - C) + D_{-1}C = I$

$D_{-q+1}(I - C) + D_{-q}C = 0$ $\qquad$ $D_1(I - C) + D_0 C = 0$

- - - - - - - $\qquad$ - - - - - - -

$D_{-1}(I - C) + D_{-2}C = 0$ $\qquad$ $D_p C = 0.$

But $(I - C)C = C(I - C)$. Therefore $D_{-q+1}(I - C)^2 = 0$ and if $D_{-q+i}(I - C)^{i-1} = 0$, then $D_{-q+i+1}(I - C) + D_{-q+i}C = 0$. Therefore, $D_{-1}(I - C)^q = 0$. Similarly $D_p C = 0$ and $D_p(I - C) + D_{p-1}C = 0$ implies $D_{p-1}C^2 = 0$. After p steps we obtain $D_0 C^p = 0$. Therefore $I = D_0(I - C) + D_{-1}C$ implies $(I - C)^q C^p =$ $D_0(I - C)^{q+1}C^p + D_{-1}C^{p+1}(I - C)^q = 0$. That is we can assume there exist r and s such that $C^r(I - C)^s = 0$ for r and s are ≥ 0.

Hence we can assume F is a free R module on n generators $g: F \longrightarrow F$ is represented by C where $C^r(I - C)^s = 0$ and $A = I + (x-1)C$. Let $P = \ker g^r$ and $Q = \ker(1-g)^s$. x^r and $(1-x)^s$ are relatively prime polynomials in Z[x]. Hence there exist polynomials f(x) and $h(x) \in Z[x]$ such that $f(x)x^r + h(x)(1-x)^s = 1$. Therefore $f(g)g^r + h(g)(1-g)^s = 1$. Therefore, if $m \in F$ then $m = g^r f(g)m + (1-g)^s h(g)Z$ and $g^r f(g)m \in Q$ and $(1-g)^s h(g)m \in P$. Hence $P + Q = F$. If $Z \in P \cap Q$ then $Z = f(g)g^r z + h(g)(1-g)^s z = 0$. Hence Hence $F = P \oplus Q$. Therefore, P and Q are finitely generated projective R modules. We claim $P(A) = Q$. Note this gives another proof of lemma 16.7.

Let g_P be g restricted to P and g_Q be g restricted to Q. By definition

$$F[X] \xrightarrow{1+(x-1)g} F[X] \longrightarrow P(A) \longrightarrow 0 \text{ is exact.}$$

This is the same as

$$\begin{array}{ccc} P[X] & \xrightarrow{1+(x-1)g_P} & P[X] \\ \oplus & & \oplus \\ Q[X] & \xrightarrow{1+(x-1)g_Q} & Q[X] \end{array}$$

since P[X] and Q[X] are stable under g. g_P is nilpotent since $P = \ker g^r$. Therefore, $1 + (x-1)g_P$ is a unipotent and hence an R[X] iso. Thus, the coker of $1 + (x-1)g_P = 0$. $(1_Q - g_Q)^s = 0$ and $1 + (x - 1)g_Q = 1 - g_Q + xg_Q$. Therefore g_Q is a unipotent and hence an iso. Therefore, we can replace $1 - g_Q + xg_Q$ by $x1_Q + (1 - g_Q)g_Q^{-1}$ since the coker will be the same. But this is exactly the set up in lemma 16.9.d). Hence, Q is the coker of $1 + (x - 1)g_Q$, and $Q = P(A)$ as claimed.

We return to $A \in GL(R)$. $[A] = [F[X, X^{-1}], 1 + (x - 1)g] = [P[X, X^{-1}], 1 + (x - 1)g_P] + [Q[X, X^{-1}], 1 + (x - 1)g_Q]$ where g_P is nilpotent and $1 - g_Q$ is nilpotent.
$[P[X, X^{-1}], 1 + (x-1)g_P] + [Q[X, X^{-1}], 1] = [F[X, X^{-1}], 1 + (x-1)h]$ where h restricted to P is g_P and h restricted to Q is 0. Then the Q part is equal to 0 in $K_1(R[X, X^{-1}])$ and h is nilpotent with entries in R. Therefore $[P[X, X^{-1}], 1 + (x - 1)g_P] \in W$. Finally $[Q[X, X^{-1}], 1 + (x - 1)g_Q] = [Q[X, X^{-1}], 1 - g_Q + xg_Q]$, and $1 - g_Q = m$ is nilpotent. But $1 - g_Q + xg_Q = m + x(1 - m) = x(1 - m + x^{-1}m) = x(1 + (x^{-1} - 1)m)$. Therefore

$[Q[X, X^{-1}], 1 + (x - 1)g_Q] = [Q[X, X^{-1}], X] + [Q[X, X^{-1}], 1 + (X^{-1}-1)m]$. But $[Q[X, X^{-1}], X]$ is in the image of $K_0(R)$ and $[Q[X, X^{-1}], 1 + (X^{-1}-1)m]$ is in w by the argument above. Therefore $[A] \in K_0(R) \oplus K_1(R) \oplus W$.

Done with lemma 16.10 and theorem 16.4.

Now we return to proving theorem 16.2. We prove a stronger theorem.

Definition. Let R be a noetherian ring. $\underline{G_1(R)} = K_1$(category of finitely generated R modules).

There is a map $K_1(R) \longrightarrow G_1(R)$ induced by inclusion of categories.

Theorem 16.11. Let R be a left regular ring. Then the map $K_1(R) \longrightarrow G_1(R)$ is an iso.

Remark. This implies theorem 16.2. If $[(P, f)] \in K_1(R)$ where $f = I + N$ is a unipotent, then in $G_1(R)$ we have

$[(P, f)] = \sum_{i=0}^{n}[N^iP/N^{i+1}P, 1] = 0$ where $N^n = 0$. Clearly f induces 1 on $N^iP/N^{i+1}P$. If R is a Dedekind ring, then submodules of projective modules are projective and theorem 16.2 can be proved without using $G_1(R)$. The difficulty is that (P, f) is not a projective object if P is. We need to lift automorphisms to resolutions and compare liftings.

Theorem 16.12. Let $\underline{A}$ be an abelian category and let $\underline{P} \subset \underline{M} \subset \underline{A}$ be full subcategories such that

1) $\underline{P}$ and $\underline{M}$ are closed under finite direct sums and $0 \in \underline{P}$,
2) If $0 \longrightarrow A' \longrightarrow A \longrightarrow A'' \longrightarrow 0$ is exact in $\underline{A}$, then A and $A'' \in \underline{M}$ implies $A' \in$ M and A and $A'' \in \underline{P}$ implies $A' \in \underline{P}$, and

3) If $M \in \underline{M}$, then there exists an exact sequence

$0 \to P_n \to \dots \to P_0 \to M \to 0$ with all the $P_i \in \underline{P}$

(n can depend on M).

Then $i: K_0(\underline{P}) \to K_0(\underline{M})$ induced by the inclusion is an iso.

<u>Proof</u>. First we show the map is onto. Pick $M \in \underline{M}$ and $0 \to P_n \to \dots \to P_0 \to M \to 0$ as in 3). Then $\sum(-1)^i[P_i]$ goes to M. For $0 \to K \to P_0 \to M \to 0$ is exact. P and $M \in \underline{M}$ implies $K \in \underline{M}$. $[M] = -[K] + [P_0]$ in $K_0(\underline{M})$. But $[K] = \sum_{i=1}^{n}(-1)^{i-1}[P_i]$ by induction. Therefore $[M] = \sum_{i=0}^{n}(-1)^i]P_i]$ as claimed and the map is onto.

Next we want to construct a $g: K_0(M) \to K_0(P)$ such that $gi = 1_{K_0(\underline{P})}$. Pick $M \in M$ and pick a resolution $0 \to P_n \to \dots \to P_0 \to M \to 0$ for M as in 3). Then we define $g([M]) = \sum_{i=0}^{n}(-1)^i[P_i]$. We must show that g is well defined and that if $0 \to M' \to M \to M'' \to 0$ is exact with M', M, and $M'' \in \underline{M}$ then $g([M]) = g([M']) + g([M''])$.

We introduce the following piece of notation. $P. \to M$ means that P. is a finite resolution for M as in 3).

$$\begin{array}{ccc} P. & \longrightarrow & M \\ \downarrow & & \downarrow \\ Q. & \longrightarrow & N \end{array}$$

means we have maps $P_i \to Q_i$ making the entire diagram commute. Also let $[P.] = \sum(-1)^i[P_i]$.

The difficulty is if we have $P. \to M$ and $Q. \to M$ there may be no map $P. \to Q.$ lifting $1: M \to M$.

Lemma 16.13. Given $Q. \longrightarrow N$ and $f\colon M \longrightarrow N$ there exists $P. \longrightarrow M$ and $P. \longrightarrow Q.$ such that

$$\begin{array}{ccc} P. & \longrightarrow & M \\ \downarrow & & \downarrow f \\ Q. & \longrightarrow & N \end{array}$$

commutes.

Proof.

Let

$$\begin{array}{ccc} X & \longrightarrow & M \\ \downarrow & & \downarrow \\ Q_0 & \longrightarrow & N \end{array}$$

be a pullback diagram.

Then there is an exact sequence

$$0 \longrightarrow X \longrightarrow M \oplus Q_0 \longrightarrow N \longrightarrow 0.$$

$M \in \underline{M}$ and $Q_0 \in \underline{M}$. Therefore $M \oplus Q_0 \in \underline{M}$. $M \oplus Q_0 \in \underline{M}$ and $N \in \underline{M}$ implies $X \in M$. Therefore, there is a $P_0 \in \underline{P}$ and an epi $P_0 \longrightarrow X$. But $X \longrightarrow M$ is an epi since $Q_0 \longrightarrow N$ is. Therefore the diagram

$$\begin{array}{ccccc} P_0 & \longrightarrow & M & \longrightarrow & 0 \\ \downarrow & & \downarrow f & & \\ Q_0 & \longrightarrow & N & \longrightarrow & 0 \end{array}$$

commutes.

Suppose we have lifted f to

$$\begin{array}{ccccc} & & P_n & \xrightarrow{d} & P_{n-1} \\ & & \downarrow & & \downarrow \\ Q_{n+1} & \longrightarrow & Q_n & \xrightarrow{d'} & Q_{n-1} \end{array}$$

such that $P_n \longrightarrow P_{n-1} \longrightarrow \dots \longrightarrow P_0 \longrightarrow M \longrightarrow 0$ is exact.

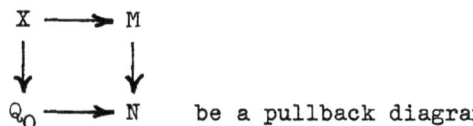

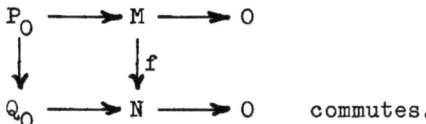

Let

$$\begin{array}{ccc} X & \longrightarrow & \ker d \\ \downarrow & & \downarrow \\ Q_{n+1} & \longrightarrow & \ker d' \longrightarrow 0 \end{array} \quad \text{be a}$$

pullback diagram. $0 \longrightarrow \ker d \longrightarrow P_n \longrightarrow \text{im } d \longrightarrow 0$ is exact. $\text{im } d = \ker (P_{n-1} \longrightarrow P_{n-2})$. Therefore $\text{im } d \in \underline{M}$ by induction on n. Therefore, $\ker d \in \underline{M}$ by a). Therefore $X \in M$ as above and there is an epi $P_{n+1} \longrightarrow X$ with $P_{n+1} \in \underline{P}$. The diagram

$$\begin{array}{ccccc} P_{n+1} & \longrightarrow & P_n & \xrightarrow{d} & P_{n-1} \\ \downarrow & & \downarrow & & \downarrow \\ Q_{n+1} & \longrightarrow & Q_n & \longrightarrow & Q_{n-1} \end{array} \quad \text{commutes.}$$

Let Q. stop at N. Then we have

$$\begin{array}{ccccccccccc} & & & & & & P_N & \xrightarrow{d} & P_{N-1} & \longrightarrow \\ & & & & & & \downarrow & & \downarrow & \\ \ldots 0 & \longrightarrow & 0 & \longrightarrow & 0 & \longrightarrow & Q_N & \longrightarrow & Q_{N-1} & \longrightarrow \end{array}$$

Let K be ker d. Then $K \in \underline{M}$. Hence there exists a resolution $0 \longrightarrow P_{N+m} \longrightarrow \ldots \longrightarrow P_{N+1} \longrightarrow K \longrightarrow 0$ for K. Hence

$$\begin{array}{ccccccccccccccc} 0 & \longrightarrow & P_{N+m} & \longrightarrow & \ldots & \longrightarrow & P_{N+1} & \longrightarrow & P_N & \longrightarrow \ldots \longrightarrow & P_0 & \longrightarrow & M & \longrightarrow & 0 \\ & & \downarrow & & & & \downarrow & & \downarrow & & \downarrow & & \downarrow f & & \\ 0 & \longrightarrow & 0 & \longrightarrow & \ldots & \longrightarrow & 0 & \longrightarrow & Q_N & \longrightarrow \ldots \longrightarrow & Q_0 & \longrightarrow & N & \longrightarrow & \end{array}$$

commutes. Done with lemma 16.13.

<u>Corollary 16.14</u>. Let $M \in M$ and $P. \longrightarrow M$ and $Q. \longrightarrow M$ be finite resolutions of M by objects of $\underline{P}$. Then there exists a

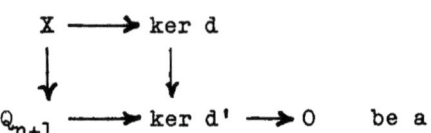

resolution $S. \longrightarrow M$ and maps $S. \longrightarrow P.$ and $S. \longrightarrow Q.$ such that

$$\begin{array}{ccc} & P. & \\ \nearrow & & \searrow \\ S. & \longrightarrow & M \\ \searrow & & \nearrow \\ & Q. & \end{array}$$

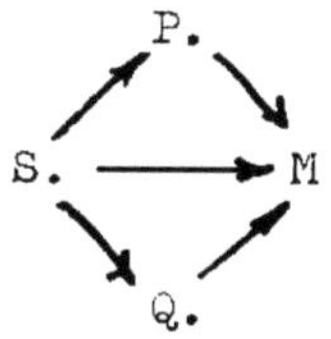

commutes.

Proof. $P. \oplus Q.$ is a resolution for $M \oplus M$ since $\underline{P}$ is closed under finite direct sums. Let $D: M \longrightarrow M \oplus M$ be the diagonal. Apply lemma 16. to obtain

$$\begin{array}{ccc} S. & \longrightarrow & M \\ \downarrow & & \downarrow \\ P. \oplus Q. & \longrightarrow & M \oplus M. \end{array}$$

Then projecting on the first factor gives $S. \longrightarrow P.$ and on the second gives $S. \longrightarrow Q.$

Done with corollary 16.14.

Now we discuss the mapping cone. Let C. and D. be two chain complexes and let $f: C. \longrightarrow D.$ be a map of chain complexes. We define a new chain complex T. by $T_n = C_{n-1} \oplus D_n$ with $d_T = T_{n+1} \longrightarrow T_n$ given by the diagram

$$\begin{array}{ccccc} C_{n-1} & \oplus & D_n & = & T_n \\ {\scriptstyle -d_C}\uparrow & \overset{f}{\nearrow} & \uparrow{\scriptstyle d_D} & & \\ C_n & \oplus & D_{n+1} & = & T_{n+1} \end{array}$$

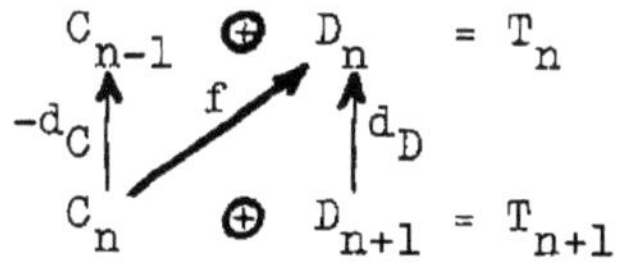

$d_D f = f d_C$ therefore T. is a chain complex. T. has D. as a subcomplex where the quotient is C. with dimension shifted down one. The sequence $0 \longrightarrow D. \longrightarrow T. \longrightarrow C. \longrightarrow 0$ is exact. Therefore, we have the long exact sequence in homotopy:

$$\to H_{n+1}(D.) \to H_{n+1}(T.) \to H_n(C.) \xrightarrow{\delta} H_n(D.) \to H_n(T.) \to$$

and $\delta = H_n(f)$ by an easy calculation. Therefore, H(f) is an iso for every dimension if and only if H(T.) = 0.

Let P. be a finite chain complex in $\underline{P}$. Define $[P.] = \sum(-1)^i[D_i] \in K_0(P)$. Let Q. be another chain complex in $\underline{P}$ and f: P. $\to$ Q.. Then [T.] = [Q.] - [P.] where T. is the mapping cone.

We claim that if P. is a finite chain complex in $\underline{P}$, then (P.) = 0 implies [P.] = 0. This is certainly true if P. has length 0. We use induction on the number of terms in P.

By assumption $0 \to P_n \to \dots \to P_m \to 0$ is exact. We break this up into

$$0 \to P_n \to \dots \to P_{m+2} \to K \to 0 \text{ and } 0 \to K \to P_{m+1} \to P_m \to 0.$$

Then $K \in \underline{P}$ by 2). Therefore $[P_{n+1}] = [P_m] + [K]$ and

$$\sum_{i=m+2}^{n}(-1)^i[P_i] + (-1)^{m+1}[K] = 0$$ using the inductive hypothesis.

Therefore [P.] = 0 + 0 = 0.

Now we can show g: $K_0(\underline{M}) \to K_0(\underline{P})$ is well defined. Let P. $\to$ M and Q. $\to$ M be resolutions of M. By corollary 16.14 there is a resolution S. $\to$ M with S. $\xrightarrow{f}$ P. and S. $\xrightarrow{h}$ Q.. Let T. be the mapping cone of f. Then $H_n(S.) = H_n(P.) = \begin{cases} 0 & n \neq 0 \\ M & n = 0 \end{cases}$, and H(f) is an iso in homology. Therefore H(T.) = 0 and [T.] = 0.

But $[T.] = [P.] - [S.]$ so $[P.] = [S.]$. Similarly $[Q.] = [S.]$. Therefore, $[P.] = [Q.]$ and g is well defined.

Next we show that if $0 \longrightarrow M' \xrightarrow{i} M \longrightarrow M'' \longrightarrow 0$ is exact that $g([M]) = g([M']) + g([M''])$. Choose a resolution $P. \longrightarrow M$ and consider

$$\begin{array}{ccc} & & P. \\ & & \downarrow \\ M' & \longrightarrow & M. \end{array}$$

By lemma 16.13 there exists $P.' \xrightarrow{f} P.$ where $P.'$ is a resolution for M'. Let $C.$ be the mapping cone of f. Then $C_n = P'_{n-1} \oplus P_n$ is a finite complex in $\underline{P}$. The sequence

$$0 = H_2(P.) \to H_2(C.) \to H_1(P.') \to H_1(P.) \to H_1(C.) \to H_0(P.') \to H_0(P.) \to H_0(C.) \to 0$$

is exact. Therefore, $H_n(C.) = 0$ if $n \geq 2$. $M' = H_0(P.') \xrightarrow{H_0(i)} H_0(P.) = M$ is a mono. Therefore, $H_1(C.) = 0$, and $H_0(C.) = \operatorname{coker} i = M''$. Therefore, $C.$ is a finite resolution in $\underline{P}$ of M''. But $[C.] = [P.] - [P.]$. Therefore, $g([M]) = g([M']) + g([M''])$.

It is clear that g is the inverse of $K_0(\underline{P}) \longrightarrow K_0(\underline{M})$.

Done with theorem 16.12.

<u>Theorem 16.15</u>. Let $\underline{A}$ be an abelian category and $\underline{P} \subset \underline{M} \subset \underline{A}$ be full subcategories such that

1) $\underline{P}$ and $\underline{M}$ are closed under finite direct sums, and $0 \in \underline{P}$.
2) If $0 \longrightarrow A' \longrightarrow A \longrightarrow A'' \longrightarrow 0$ is exact in $\underline{A}$, then A and $A'' \in \underline{P}$ implies $A' \in \underline{P}$ and A and $A'' \in M$ implies $A' \in \underline{M}$.
3) If $M \in \underline{M}$, there exists $P \in \underline{P}$ and an epi $P \longrightarrow M$ such that every automorphism of M lifts to an automorphism of P, and

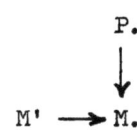

4) If $\to P_n \to P_{n-1} \to \ldots \to P_0 \to M \to 0$ is exact with $M \in \underline{M}$ and $P_i \in \underline{P}$, then there is an n (depending on the resolution) such that $\ker[P_n \to P_{n-1}] \in \underline{P}$. Then the inclusion $\underline{P} \subset \underline{M}$ induces an iso $K_1(\underline{P}) \to K_1(\underline{M})$.

<u>Proof</u>. $K_1(\underline{P}) = K_0(\underline{P}[t, t^{-1}])/[(P, fg)] - [(P, f)] - [(P, g)]$.

We first show that theorem 16.12 applies to $\underline{\underline{P}[X, X^{-1}]} \subset \underline{\underline{M}[X, X^{-1}]}$. Conditions (1) and (2) are immediate. For (3), we construct a resolution of M as follows. By hypothesis (3), we can find $P_0 \in \underline{P}$ and $P_0 \to M \to 0$ so any automorphism of M lifts to one of P_0. Repeat this process on $\ker P_0 \to M$, etc. This gives a resolution $\ldots \to P_n \to P_{n-1} \to \ldots \to P_0 \to M \to 0$ such that any automorphism of M lifts to one of P. Now by hypothesis (4), there is an n such that $P'_{n+1} = \ker[\underline{P}_n \to P_{n-1}] \in \underline{P}$. Replace the resolution by $0 \to P'_{n+1} \to P_n \to P_{n-1} \to \ldots \to P_0 \to M \to 0$. This shows that there is a <u>finite</u> resolution P. of M by objects in $\underline{P}$ such that every automorphism of M lifts to one of P. Now if $(M, f) \in \underline{\underline{M}[X, X^{-1}]}$, lift f to an automorphism f. of P.. Then (P., f.) is a resolution of (M, f) in $\underline{\underline{P}[X, X^{-1}]}$. Now by theorem 16.12, $K_0(\underline{\underline{P}[X, X^{-1}]}) \to K_0(\underline{\underline{M}[X, X^{-1}]})$ is an isomorphism with inverse $F: K_0(\underline{M}(X, X^{-1}]) \to K_0(\underline{\underline{P}[X, X^{-1}]})$ given by $F[(M, f)] = [(P., f.)]$. To get K_1 we must factor out the subgroups generated by all $[(P, fg)] - [(P, f)] - [(P, g)]$ and $[(M, fg)] - [(M, f)] - [(M, g)]$. To get an inverse for $K_1(\underline{P}) \to K_1(\underline{M})$ it will suffice to show that F preserves these subgroups and take the map $K_1(\underline{M}) \to K_1(\underline{P})$ induced by F. Now, using the resolution for M constructed above, we can lift the automorphisms f, g of M to f., g. on P. Then f.g. lifts fg so F sends [(M, fg)] - [(M, f)] - [(M, g)] to [(P., f.g.)] - [(P., f.)] - [(P., g.)] which goes to 0 in $K_1(\underline{P})$.

Done.

We now give a sufficient condition for hypothesis (3) of this theorem to hold.

Proposition 16.16. Let $\underline{A}$ be an abelian category and let $\underline{P} \subset \underline{M} \subset \underline{A}$ be full subcategories each closed under finite direct sums. Suppose for each $M \in \underline{M}$, there is an epimorphism $P \twoheadrightarrow M$ with $P \in \underline{P}$ such that every endomorphism of M lifts to an endomorphism of P. Then there is an epimorphism $Q \twoheadrightarrow M$ with $Q \in \underline{P}$ such that every automorphism of M lifts to an automorphism of Q.

Proof. Let $Q = P \oplus P$ and map $Q \twoheadrightarrow M$ by $P \oplus P \twoheadrightarrow M \oplus M \xrightarrow[pr_1]{} M$. Let h be an automorphism of M. Lift h to the automorphism $h \oplus h^{-1}$ of $M \oplus M$. In matrix form $h \oplus h^{-1} = \begin{pmatrix} h & 0 \\ 0 & h^{-1} \end{pmatrix}$. But this is a product of elementary matrices of the form $\begin{pmatrix} 1 & f \\ 0 & 1 \end{pmatrix}$ and $\begin{pmatrix} 1 & 0 \\ g & 1 \end{pmatrix}$. Now $\begin{pmatrix} 1 & f \\ 0 & 1 \end{pmatrix}$ is given by the diagram

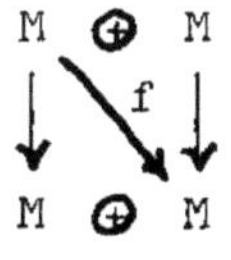

Lift f to an endomorphism $f'\colon P \longrightarrow P$. Then $\begin{pmatrix} 1 & f \\ 0 & 1 \end{pmatrix}$ lifts to the elementary automorphism $\begin{pmatrix} 1 & f' \\ 0 & 1 \end{pmatrix}$ of $P \oplus P$. Treating each elementary factor in this way and taking the product, we lift $\begin{pmatrix} h & 0 \\ 0 & h^{-1} \end{pmatrix}$ to an automorphism of $P \oplus P = Q$.

Done.

In particular, this result holds if $\underline{M}$ = all finitely generated R-modules and $\underline{P}$ = all finitely generated projective R-modules. If R is also left noetherian, these categories satisfy (2) of theorem 16.15 and if R is left regular, they also satisfy (4). This proves theorem 15.2. More generally

Corollary 16.17. If R is left noetherian then $K_1(R) \approx K_1$(category of finitely generated R-modules of finite projective dimension).

It is possible to prove corollary 16.17 by a simpler method which, however, does not lead to theorems as general as those above. We will state the result for modules although the generalization to abelian categories is immediate.

Theorem 16.18. Let R be left noetherian and let $\underline{M}$ be the category of finitely generated left R modules. Let $\underline{A} \subset \underline{B} \subset \underline{M}$ be full subcategories such that $\underline{A}$ and $\underline{B}$ are closed under finite direct sums and contain 0. Assume that if $0 \to X \to P \to B \to 0$ in $\underline{B}$ and P is projective, then $X \in \underline{A}$. Then $K_0(\underline{A}) \to K_0(\underline{B})$ and $K_1(\underline{A}) \to K_1(\underline{B})$ are isomorphisms.

Proof. We will give the proof for K_1. The proof for K_0 is obtained in the same way ignoring all automorphisms. Let $B \in \underline{B}$ and let b be an automorphism of B. We want to define $F(B, b) \in K_1(\underline{A})$. By proposition 16.16 we can find a finitely generated projective P and an epimorphism $P \to B$ such that all automorphisms of B lift to P. Lift b to P and get

$$0 \to (X, x) \to (P, p) \to (B, b) \to 0.$$

Define $F(B, b) = [(P, p)] - [(X, x)]$. This makes sense because

$P \in \underline{A}$ by the hypothesis applied to $0 \to P \to P \to 0 \to 0$. We will show that F gives a well defined map $K_1(\underline{B}) \to K_1(\underline{A})$. It is then clear that this will be an inverse for $K_1(\underline{A}) \to K_1(\underline{B})$. Suppose we make a different choice

$$0 \to (X', x') \to (P', p') \to (B, b) \to 0.$$

Form the pullback diagram

$$\begin{array}{ccccccccc} & & & & 0 & & & & \\ & & & & \downarrow & & & & \\ & & & & (X, x) & & & & \\ & & & & \downarrow & & & & \\ 0 & \to & (X', x') & \to & (M, m) & \to & (P, p) & \to & 0 \\ & & & & \downarrow & & \downarrow & & \\ & & & & (P', p') & \to & (B, b) & & \\ & & & & \downarrow & & & & \\ & & & & 0 & & & & \end{array}$$

Since P' is projective the sequence $0 \to X \to M \to P' \to 0$ splits. Therefore $M \cong P' \oplus X$ so $M \in \underline{A}$. Since $[(M, m)] = [(P, p)] + [(X', x')] = [(P', p')] + [(X, x)]$ we see that the two definitions of $F(B, b)$ agree.

We now check the relations defining $K_1(\underline{B})$. If b, b' are automorphisms of B, lift them to p, p' on P getting x, x' on X. Thus $F(B, bb') = [(P, pp')] - [(X, xx')] = F(B, b) + F(B, b')$ using the relation in $K_1(\underline{A})$. Finally, let $0 \to (B', b') \to (B, b) \to (B'', b'') \to 0$ be an exact sequence. Choose $P \to B$ and $P' \to B'$ such that automorphisms lift. Lift b to p, b' to p' and construct the following diagram using the compositions $(P', p') \to (B', b') \to (B, b)$ and $(P, p) \to (B, b) \to (B'', b'')$.

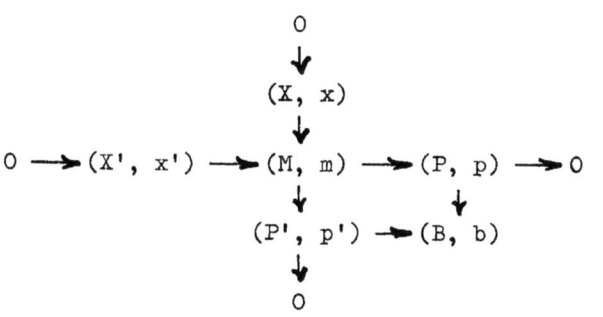

$$
\begin{array}{ccccccccc}
 & & 0 & & 0 & & 0 & & \\
 & & \downarrow & & \downarrow & & \downarrow & & \\
0 & \longrightarrow & (X', x') & \longrightarrow & (X, x) & \longrightarrow & (X'', x'') & \longrightarrow & 0 \\
 & & \downarrow & & \downarrow & & \downarrow & & \\
0 & \longrightarrow & (P', p') & \longrightarrow & (P', p') \oplus (P, p) & \longrightarrow & (P, p) & \longrightarrow & 0 \\
 & & \downarrow & & \downarrow & & \downarrow & & \\
0 & \longrightarrow & (B', b') & \longrightarrow & (B, b) & \longrightarrow & (B'', b'') & \longrightarrow & 0 \\
 & & \downarrow & & \downarrow & & \downarrow & & \\
 & & 0 & & 0 & & 0 & &
\end{array}
$$

The top row is exact by the 9-lemma. Using the 3 columns to define the F's we see that $F(B, b) = F(B', b') + F(B'', b'')$. Done.

<u>Corollary 16.19</u>. Let R be left noetherian. Let $\underline{\underline{H}}_n$ be the full subcategory of finitely generated left R modules of projective dimension $\leq n$. Then $K_1(\underline{\underline{H}}_n) \longrightarrow K_1(\underline{\underline{H}}_{n+1})$ and $K_0(\underline{\underline{H}}_n) \longrightarrow K_0(\underline{\underline{H}}_{n+1})$ are isomorphisms.

It follows that $K_1(R) \longrightarrow K_1(\underline{\underline{H}}_n)$ and $K_0(R) \longrightarrow K_0(\underline{\underline{H}}_n)$ are isomorphisms. If $\underline{\underline{H}}_\infty = \bigcup \underline{\underline{H}}_n$, then $K_i(\underline{\underline{H}}_\infty) = \varinjlim K_i(\underline{\underline{H}}_n)$ (directly from the definition) and so $K_i(R) \longrightarrow K_i(\underline{\underline{H}}_\infty)$ is an isomorphism $(i = 0, 1)$.

We can also apply the theorem to $G_1(R) = K_1(\underline{M})$. For example, in the case of an algebra Λ with $\underline{M}$ = finitely generated Λ-modules,

<u>Corollary 16.20</u>. Let R be regular and commutative. Let Λ be an R-algebra, finitely generated and projective as an R-module. Let $\underline{Q} \subset \underline{M}$ be the full sub-category of finitely generated Λ-modules which are projective over R. Then $K_i(\underline{Q}) \longrightarrow K_i(\underline{M}) = G_i(\Lambda)$ is an isomorphism $(i = 0, 1)$.

<u>Proof</u>. Let $\underline{Q}_n \subset \underline{M}$ consist of those Λ modules with projective

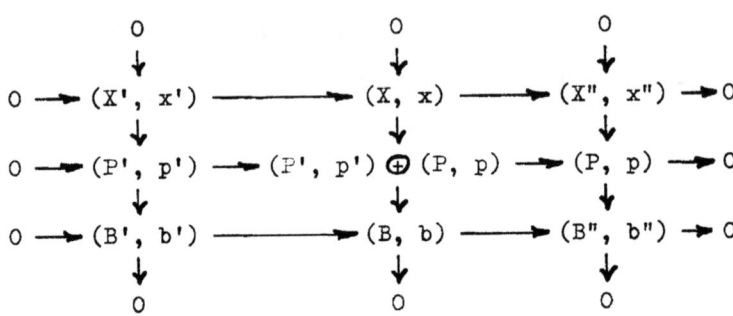

dimension $\leq$ n over R. Since R is regular, $\underline{M}_- = \bigcup \underline{Q}_n$ so $K_i(\underline{M}) = \varinjlim K_i(\underline{Q}_n)$. But $K_i(\underline{Q}_n) \to K_i(\underline{Q}_{n+1})$ is an isomorphism by theorem 16.18 (the hypothesis applies since Λ is projective over R). Since $\underline{Q} = \underline{Q}_0$ we are done.

Similarly we get

Corollary 16.21. Let R be a commutative noetherian integral domain. Let Λ be an R-algebra finitely generated and torsion free as an R-module. Let $\underline{T} \subset \underline{M}$ be the full subcategory of Λ-modules finitely generated and torsion free over R. Then $K_i(\underline{T}) \to K_i(\underline{M}) = G_i(\Lambda)$ is an isomorphism.

In this case the theorem applies directly to $\underline{T} \subset \underline{M}$ since any projective Λ-module or submodule thereof is torsion free.

Chapter 17. Relations Between Algebraic and Topological K Theory

Let X be a compact Hausdorff space and let R = C(X) be the ring of continuous real or complex functions on X. If ξ is a real or complex vector bundle over X, the group of sections $\Gamma(\xi)$ can be viewed as an R-module. It can be shown that the functor Γ establishes an equivalence between the category of vector bundles on X and the category of finitely generated projective R-modules (see R. G. Swan, Vector bundles and projective modules, Trans. Amer. Math. Soc. 105 (1962)). In particular we have $K_0(R) = K^0(X)$.

We now give an analogous result for K_1. The ring R = C(X) has an obvious topology. In fact it is a Banach algebra with $||f|| = \sup|f(x)|$. Therefore the group of units R^* is a topological group. Let R^{*0} denote its connected component.

Theorem 17.1. Let X be a compact Hausdorff space and let $R = C(X)$ be the ring of continuous real or complex functions on X. Then there is an exact sequence

$$0 \to R^{*0} \to K_1(R) \to \tilde{K}^{-1}(X) \to 0$$

where $\tilde{K}^{-1}$ is the real or complex topological $\tilde{K}^{-1}$ functor.

Proof:

Let GL be $GL(\mathbb{R})$ or $GL(\mathbb{C})$ as the case may be. Then $\tilde{K}^{-1}(X) = [X, GL]$ (Atiyah lemma 2.4.6), the group of homotopy classes of maps $X \to GL$. Now a map $X \to GL$ has its image in some $GL(n, \mathbb{R})$ or $GL(n, \mathbb{C})$ and so is represented by a $n \times n$ invertible matrix whose entries are functions on X, i.e., by an element of $GL(n, R)$. Therefore the group of maps $X \to GL$ is just $GL(R)$. Since $\tilde{K}^{-1}(X)$ is a quotient of this and is commutative, we get an epimorphism $K_1(R) \xrightarrow{\eta} \tilde{K}^{-1}(X) \to 0$. Since R^{*0} is connected, any element of R^{*0} can be deformed continuously into the unit $1 \in R$. Since $\tilde{K}^{-1}(X)$ is obtained from $GL(R)$ by identifying homotopic elements, $R^{*0} \subset \ker \eta$. We must now show the opposite inclusion. Let $a \in GL(R)$ and suppose there is a homotopy a_t, $t \in [0, 1]$ with $a_0 = a$, $a_1 = 1$. Since $X \times [0, 1]$ is compact, its image in $GL(\mathbb{R})$ or $GL(\mathbb{C})$ lies in some $GL(n, \mathbb{R})$ or $GL(n, \mathbb{C})$. Therefore a_t is represented by an $n \times n$ matrix A_t, $A_0 = A$ represents a, and $A_1 = I$ is the identity matrix. Divide [0, 1] into subintervals $0 = t_0 < t_1 < \dots < t_k = 1$ so that A_{t_i} and $A_{t_{i+1}}$ are very close. Write $A_i = A_{t_i}$ and $A = A_0 = (A_0A_1^{-1})(A_1A_2^{-1}) \dots (A_{k-1}A_k^{-1})$. Note $A_k = I$. If the subdivision is fine enough, each $A_iA_{i+1}^{-1}$ is so close to I that we can reduce it to diagonal form by elementary transformations, subtracting multiples of the principal diagonal elements from all the other entries. Therefore a is congruent mod E(R) to a diagonal matrix so the image of a in $K_1(R)$ is represented also by an element $r \in R^*$. In fact r = det a. Now $r_t = \det a_t$ gives a path from r to 1 in R^* so $r \in R^{*0}$. Therefore $\ker \eta \subset R^{*0}$. Done.

We now restrict ourselves to the complex case. The object is to prove the complex periodicity theorem using the algebraic techniques we have developed.

Theorem 17.2. Let X be a compact hausdorff space. Then, in complex K-theory, we have $\tilde{K}^{-1}(X \times S^1) \approx \tilde{K}^{-1}(X) \oplus K^0(X)$. The injection and projection of the summand $\tilde{K}^{-1}(X)$ are obtained from the maps $X \longrightarrow X \times S^1$ (by a cross section) and $X \times S^1 \longrightarrow X$ by projection.

It is very easy to deduce the periodicity theorem from this. By Atiyah Cor. 2.4.8 we have $\tilde{K}^{-1}(X \times S^1) \approx \tilde{K}^{-1}(X \cap S') \oplus \tilde{K}^{-1}(X) \oplus \tilde{K}^{-1}(S')$. Now $\tilde{K}^{-1}(S') = \mathbb{Z}$. Applying the theorem and noting that the summand $\tilde{K}^{-1}(X)$ is the same in both isomorphisms, we get $K^0(X) \approx \mathbb{Z} \oplus \tilde{K}^{-1}(X \cap S^1)$. Applying this to X = point we see that the projection $K^0X \longrightarrow \mathbb{Z}$ is the usual one $K^0X \longrightarrow K^0(\text{pt.})$ (Note that all our isomorphisms are natural.) Therefore we get $\tilde{K}^0X \approx \tilde{K}^{-1}(X \cap S^1) = \tilde{K}^{-1}(SX) = \tilde{K}^{-2}(X)$.

Corollary 17.3 (Periodicity). $\tilde{K}^0(X) \approx \tilde{K}^{-2}(X)$.

Applying this to X^+ gives $K^0(X) \approx K^{-2}(X)$.

Proof of Theorem 17.2.

Suppose F is a contravariant functor from topological spaces to sets, groups, rings, etc. (anything with elements). We say two elements a, b $\in$ F(X) are homotopic if there is some g $\in$ F(X) x I),

$I = [0, 1]$ with $F(i_0)g = a$, $F(i_1)g = b$, where $i_0, i_1 \colon X \longrightarrow X \times I$ by $i_0(x) = (x, 0)$, $i_1(x) = x, 1)$. If $f\colon X \longrightarrow Y$ and $a, b \in F(Y)$ are homotopic (notation $a \simeq b$), then $F(f)a \simeq F(f)b$. To see this consider the diagram

$$\begin{array}{ccc} X & \longrightarrow & Y \\ \downarrow\downarrow & & \downarrow\downarrow \\ X \times I & \longrightarrow & Y \times I \end{array}$$

Therefore we can define a new functor $\overline{F}(X)$ by identifying homotopic elements of $F(X)$. For example if $F(X) = GL(C(X))$ then clearly $\overline{F}(X) = \widetilde{K}^{-1}(X)$. The same result holds for $F(X) = K_1(C(X))$ since we have epimorphisms $GL(C(X)) \longrightarrow K_1(C(X)) \longrightarrow \widetilde{K}^{-1}(X)$.

It is trivial to verify that homotopy is compatible with any natural group structure on $F(X)$ and that the operation taking F to $\overline{F}$ preserves natural finite direct sum decomposition. Now apply theorem 16.4 to $R = C(X)$. If $F(X) = K_1(C(X)[x, x^{-1}])$, we get $F(X) = K_1(C(X)) \oplus K_0(C(X)) \oplus W(X)$. Here $W(X)$ is generated by elements $I + (x - 1)N$, $I + (x^{-1} - 1)N$ where N is nilpotent. These elements are all homotopic to I. In fact, if we write $N(s)$ to indicate that the entries of N are functions of $s \in X$ and define $M(s, t) = tN(s)$ for $s \in X$, $t \in I$, then $I + (x - 1)M(s, t)$ represents an element of $F(X \times I)$ which gives a homotopy between $I + (x - 1)N$ and I. We now pass to $\overline{F}$. Reducing $K_1(C(X))$ module homotopies gives $\widetilde{K}^{-1}(X)$, while $K_0(C(X)) = K^0(X)$ is already a homotopy invariant. Therefore we get

$$\overline{F}(X) = \widetilde{K}^{-1}(X) \oplus K^0(X),$$

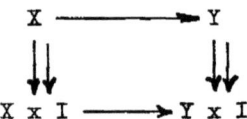

and it will suffice to show $\overline{F}(X) = \widetilde{K}^{-1}(X \times S')$. The verification that $\widetilde{K}^{-1}(X)$ is the correct summand will be trivial. The projection $X \times S' \longrightarrow X$ gives an injection $C(X) \longrightarrow C(X \times S')$. Identify S' with the unit circle in $\underline{C}$ and define $Z \in C(X \times S')$ to be $X \times S' \longrightarrow S' \longrightarrow C$. Define $C(X)[x, x^{-1}] \longrightarrow C(X \times S')$ by sending x to z. This gives $F(X) = K_1(C(X)[x, x^{-1}]) \longrightarrow K_1(C(X \times S'))$. Identifying homotopic elements gives $\overline{F}(X) \longrightarrow \widetilde{K}^{-1}(X \times S')$. It will clearly suffice to show that this is an isomorphism.

Note that $C(X)[x, x^{-1}] \longrightarrow C(X \times S')$ is injective. In fact, $\sum_{-m}^{n} f_j(s)x^j \longrightarrow \sum_{-m}^{n} f_j(s)z^j = g(s, z)$ and we can recover the $f_j(s)$ by $f_j(s) = \frac{1}{2\pi i} \int z^{-j-1} g(s, z)dz$, the integral being taken around the unit circle. Therefore we can identify $C(X)[x, x^{-1}]$ with $C(X)[z, z^{-1}] \subset C(X \times S')$ and $GL(C(X)[x, x^{-1}])$ with $GL(C(X)(z, z^{-1}]) \subset GL(C(X \times S'))$.

If F is a contravariant functor on topological spaces, $a \in F(X)$, and A is a closed subspace of X, we will write $a|A$ for $F(i)a$ where $i: A \longrightarrow X$ is the inclusion.

<u>Lemma 17.4</u>. Let X be a compact hausdorff space. Let $a \in GL(C(X \times S'))$. Then there is a $b \in GL(C(X)[z, z^{-1}])$ such that $b \simeq a$ in $GL(C(X \times S'))$. If $b_1, b_2 \in GL(C(X)[z, z^{-1}])$ and $b_1 \simeq b_2$ in $GL(C(X \times S'))$ then $b_1 \simeq b_2$ in $GL(C(X)[z, z^{-1}])$.

This lemma shows that the two functors involved give the same functor when homotopic elements are identified. Since they give $\overline{F}(X)$ and $\widetilde{K}^{-1}(X \times S')$, this will complete the proof of the theorem.

Let $H = H(X)$ be the set of matrices in $GL(C(X \times S'))$ which have all entries in $C(X)[z, z^{-1}]$. Note that H is not a subgroup. If $a \in H$, then $a \in GL(C(X)[z, z^{-1}])$ if and only if det a is a unit of $C(X)[z, z^{-1}]$. We prove the lemma in 2 stages. It is an obvious consequence of the next 2 lemmas.

Lemma 17.5. If $a \in GL(C(X \times S'))$ then there is a $b \in H$, $b \simeq a$. If $b_1, b_2 \in H$ and $b_1 \simeq b_2$ in $GL(C(X \times S'))$ then $b_1 \simeq b_2$ in H.

Proof:

The ring $C(X)[z, z^{-1}]$ separates points in $X \times S'$. Since $|z| = 1$, $\bar{z} = z^{-1}$ so this ring is stable under complex conjugation. Therefore the Stone-Weierstrass theorem shows it is dense in $C(X \times S')$. Now if $a, b \in GL(C(X \times S'))$ are sufficiently close then $a \simeq a$. In fact we need only consider the matrix $g \in GL(C(X \times I \times S'))$ defined by $g = (1 - t)a + tb$ where $t \in I$. This will be invertible if and only if det $g \neq 0$ at all points of $X \times I \times S'$ but if a and b are very close, det g will be close to det a for all $t \in I$. Suppose now $a \in GL(C(X \times S'))$ is given. By the Stone-Weierstrass theorem we find a b with entries in $C(X)[z, z^{-1}]$ very close to a. If it is close enough, det b is very close to det a and so is never 0 on $X \times S'$. Thus $b \in H$ and $b \simeq a$ if b is close enough to a. Suppose now $b_1, b_2 \in H$ and $b_1 \simeq b_2$ in $GL(C(X \times S'))$. Let $g \in GL(C(X \times I \times S'))$ be a homotopy between them. Let $g_t \in GL(C(X \times S'))$ be the image of g under the map induced by $X \times S' \to X \times I \times S'$ sending $(u, z) \to (u, t, z)$. Then $g_0 = b_1$, $g_1 = b_2$. By the first part of the proof we can find $d \in H(X \times I)$ very close to g. Therefore d_0, d_1 are very close to b_1 and b_2. Since $d_0 \simeq d_1$ in H(X) by definition, we only need to show $b_1 \simeq d_0$, $b_2 \simeq d_1$ in H(X). Suppose b, d $\in$ H(X) and are very close. Then the homotopy $(1 - t)b + td$ between them in $GL(C(X \times S'))$ will clearly lie in H(X).

Done with lemma 17.5.

Lemma 17.6. If $a \in H(X)$, there is a $b \in GL(C(X)[z, z^{-1}])$ such that $b \simeq a$ in H(X). If $b_1, b_2 \in GL(C(X)[z, z^{-1}])$ and $b_1 \simeq b_2$ in H(X), then $b_1 \simeq b_2$ in $GL(C(X)[z, z^{-1}])$.

Proof: As in the proof of lemma 16.10 we can find a diagonal matrix d with entries powers of z and two elementary matrices e, e' such that edae' is represented by a matrix $A_0 + zA_1$ with A_0, A_1 having entries in C(X). Now $\det(A_0 + zA_1)$ is never 0 on X x S' so $\det(A_0 + A_1)$ is never 0 on X. Therefore $A_0 + A_1$ represents $h \in GL(C(X))$ and $h^{-1}edae'$ is represented by a matrix of the form $I + (x - 1)C$. The remainder of the argument in lemma 16.10 will not work here because $I + (z - 1)C$ may not have an inverse in $GL(C(X)[z, z^{-1}])$. The next two lemmas make up for this.

Lemma 17.7. If $M \in M_n(\underline{C})$ then $I + (z - 1)M$ is invertible for all $|z| = 1$ if and only if all eigenvalues ξ of M satisfy $R\xi \neq 1/2$.

Proof.

The eigenvalues of $I + (z - 1)M$ are $1 + (z - 1)\xi$. These are $\neq 0$ for all $|z| = 1$ if and only if the $\xi_i \notin \{\frac{1}{1-z} \mid |z| = 1\}$. But this set is the line $R\xi = 1/2$.

Done with lemma 17.7.

Now let U be the set of $M \in M_n(\underline{C})$ defined by the condition of lemma 17.7, let V be the set of M with all eigenvalues 0 or 1 and let W be the set of idempotent matrices. Then $U \supset V \supset W$.

Lemma 17.8. There is a deformation retraction of U on W under which V is stable.

Proof.

If $M \in U$, there is a unique decomposition $\underline{C}^n = L_1 \oplus L_2$ where L_1, L_2 are stable under M, $M_1 = M|L_1$ has eigenvalues satisfying $R\,\xi < 1/2$ and $M_2 = M|L_2$ has eigenvalues satisfying $R\,\xi < 1/2$. Define the homotopy by taking $(1 - t)M_1$ on L_1 and $tI + (1 - t)M_2$ on L_2. This gives M for $t = 0$ and gives the projection on L_2 for $t = 1$. The eigenvalues of $(1 - t)M_1$ and $tI + (1 - t)M_2$ are $(1 - t)\xi$, $t + (1 - t)\xi$ respectively. Therefore the deformation never leaves U and V is stable. Clearly W is fixed.

Done with lemma 17.8.

We now return to the proof of lemma 17.6. By what we have done so far, it will suffice to prove the first part for an a represented by a matrix $I + (z - 1)C$ where $C: X \longrightarrow U$. The proof of lemma 16.10 shows that $I + (z - 1)C$ lies in $GL(C(X)[z, z^{-1}])$ if and only if C satisfies an equation $C^r(I - C)^s = 0$. This implies $C: X \longrightarrow V$ and, conversely, if $C: X \longrightarrow V$ then C satisfies $C^n(I - C)^n = 0$. Therefore it is sufficient to deform $C: X \longrightarrow U$ into some $C': X \longrightarrow V$. This is possible by lemma 17.8.

Finally let b_1, b_2 be homotopic in H(X) and let $g \in H(X \times I)$ be a homotopy between them. As in the first part we can find h^{-1}edge' represented by $I + (z - 1)C$, $C: X \times I \longrightarrow U$. We want to change this into $C': X \times I \longrightarrow V$ without changing $g_0 = b_1$ and $g_1 = b_2$. Since $b_1, b_2 \in GL(C(X)[z, z^{-1}])$ we see that $C_0, C_1: X \longrightarrow V$. We must change C into C' without changing

$C|X \times \{0\}$ and $C|X \times \{1\}$. Let $f_s : U \to U$ be the deformation of lemma 17.8 so that $f_0 = \mathrm{id}$, $f_1 : U \to W$. Define $C' : X \times I \to L$ by

$$C'(x, t) = \begin{cases} f_{3t}C(x, 0) & \text{if } 0 \leq t \leq 1/3 \\ f_1 C(x, 3t-1) & \text{if } 1/3 \leq t \leq 2/3 \\ f_{3-3t}C(x, 1) & \text{if } 2/3 \leq t \leq 1 \end{cases}$$

Since $C(x, 0)$, $C(x, 1) \in V$, this lies in V. It is clearly continuous. This completes the proof of lemma 17.6 and theorem 17.2.

Remark. There is an algebraic analogue of the homotopy relation considered above. Let F be a covariant functor on the category of rings. If a, b $\in F(R)$ define a $\simeq$ b if there is some g $\in F(R[t])$ such that $F(p_0)g' = a$, $F(p_1)g = b$ where $p_0, p_1 : R[t] \to R$ are the identity on R and $p_0(t) = 0$, $p_1(t) = 1$. Let $\simeq$ be the equivalence relation generated by this and let $\overline{F}(R) = F(R)/\simeq$ i.e., $\underline{F}(R)$ is the difference cokernel of $F(R[t]) \rightrightarrows F(R)$.

If we apply this to K_1, K_0 we get $\overline{K}_1(R)$, $\overline{K}_0(R)$.

Proposition 17.9. $\overline{K}_1(R) = GL(R)/U(R)$ where $U(R)$ is the subgroup of $K_1(R)$ generated by all unipotent elements.

Proof.

Suppose $U = I + N$ is unipotent, i.e., N is nilpotent. Then $I + tN$ represents an element of $K_1(R[t])$ whose two images are I and $I + N = U$. Thus U represents 0 in $\overline{K}_1(R)$. Let $K_1'(R) = GL(R)/U(R) = K_1(R)/\text{unipotents}$. By theorem 16.1 the map $K_1'(R) \to K_1'(R[t])$ is onto since V is generated by unipotents. Both compositions $K_1'(R) \to K_1'(R[t]) \rightrightarrows K_1'(R)$ are the identity.

Therefore $K_1'(p_0) = K_1'(p_1)$ so $\overline{K}_1'(R) = K_1'(R)$. But $K_1(R) \twoheadrightarrow K_1'(R) \twoheadrightarrow \overline{K}_1(R)$ so $\overline{K}_1(R) = \overline{K}_1'(R)$.

Corollary 17.10. For any R, $\overline{K}_1(R[t]) = \overline{K}_1(R)$ and

$$\overline{K}_1(R[t, t^{-1}]) = \overline{K}_1(R) \oplus \overline{K}_0(R).$$

This follows from theorems 16.1 and 16.4 by reducing modulo the relation $\simeq$. Conversely it is clear that $K_1(R) = K_1(R[t])$ if and only if $\overline{K}_1(R) = K_1(R)$ and $\overline{K}_1(R[t]) = K_1(R[t])$.

BIBLIOGRAPHY

M. Atiyah. K-Theory, Benjamin, New York, 1967.

H. Bass. K-Theory and Stable Algebra. Publ. Math. IHES, no. 22, Paris, 1964.

H. Bass. Algebraic K-Theory. notes Tata Institute.

H. Bass, A. Heller, and R. G. Swan, The Whitehead Group of a Polynomial Extension, Publ. Math. IHES, no. 22, Paris, 1964.

P. Freyd. Abelian Categories, Harper and Row, New York, 1964.

P. Gabriel. Des Categories Abeliennes, Bull. Soc. Math. France, 90, 1962, 323-448.

A. Heller. Exact Sequences in Algebraic·K Theory, Topology, 4, 1965, 389-408.

R. G. Swan. Induced Representations and Projective Modules, Ann. of Math., 71, 1960, 552-557.

R. G. Swan. Vector Bundles and Projective Modules, Trans. Amer. Math. Soc., 105, 1962, 264-277.

J. H. C. Whitehead. Simple Homotopy Types, Amer. J. Math., 78, 1950, 1-57.

INDEX

LIST OF SYMBOLS

Offsetdruck: Julius Beltz, Weinheim/Bergstr.

Lecture Notes in Mathematics

Bisher erschienen/Already published

Vol. 1: J. Wermer, Seminar über Funktionen-Algebren. IV, 30 Seiten. 1964. DM 3,80 / 0.95

Vol. 2: A. Borel, Cohomologie des espaces localement compacts d'après J. Leray. IV, 93 pages. 1964. DM 9,– / $ 2.25

Vol. 3: J. F. Adams, Stable Homotopy Theory. 2nd. revised edition. IV, 78 pages. 1966. DM 7,80 / $ 1.95

Vol. 4: M. Arkowitz and C. R. Curjel, Groups of Homotopy Classes. 2nd. revised edition. IV, 36 pages. 1967. DM 4,80 / $ 1.20

Vol. 5: J.-P. Serre, Cohomologie Galoisienne. Troisième édition. VIII, 214 pages. 1965. DM 18,– / $ 4.50

Vol. 6: H. Hermes, Eine Termlogik mit Auswahloperator. IV, 42 Seiten. 1965. DM 5,80 / $ 1.45

Vol. 7: Ph. Tondeur, Introduction to Lie Groups and Transformation Groups. VIII, 176 pages. 1965. DM 13,50 / $ 3.40

Vol. 8: G. Fichera, Linear Elliptic Differential Systems and Eigenvalue Problems. IV, 176 pages. 1965. DM 13,50 / $ 3.40

Vol. 9: P. L. Ivănescu, Pseudo-Boolean Programming and Applications. IV, 50 pages. 1965. DM 4,80 / $ 1.20

Vol. 10: H. Lüneburg, Die Suzukigruppen und ihre Geometrien. VI, 111 Seiten. 1965. DM 8,– / $ 2.00

Vol. 11: J.-P. Serre, Algèbre Locale. Multiplicités. Rédigé par P. Gabriel. Seconde édition. VIII, 192 pages. 1965. DM 12,– / $ 3.00

Vol. 12: A. Dold, Halbexakte Homotopiefunktoren. II, 157 Seiten. 1966. DM 12,– / $ 3.00

Vol. 13: E. Thomas, Seminar on Fiber Spaces. IV, 45 pages. 1966. DM 4,80 / $ 1.20

Vol. 14: H. Werner, Vorlesung über Approximationstheorie. IV, 184 Seiten und 12 Seiten Anhang. 1966. DM 14,– / $ 3.50

Vol. 15: F. Oort, Commutative Group Schemes. VI, 133 pages. 1966. DM 9,80 / $ 2.45

Vol. 16: J. Pfanzagl and W. Pierlo, Compact Systems of Sets. IV, 48 pages. 1966. DM 5,80 / $ 1.45

Vol. 17: C. Müller, Spherical Harmonics. IV, 46 pages. 1966. DM 5,– / $ 1.25

Vol 18: H.-B. Brinkmann und D. Puppe, Kategorien und Funktoren. XII, 107 Seiten, 1966. DM 8,– / $ 2.00

Vol. 19: G. Stolzenberg, Volumes, Limits and Extensions of Analytic Varieties. IV, 45 pages. 1966. DM 5,40 / $ 1.35

Vol. 20: R. Hartshorne, Residues and Duality. VIII, 423 pages. 1966. DM 20,– / $ 5.00

Vol. 21: Seminar on Complex Multiplication. By A. Borel, S. Chowla, C. S. Herz, K. Iwasawa, J.-P. Serre. IV, 102 pages. 1966. DM 8,– / $ 2.00

Vol. 22: H. Bauer, Harmonische Räume und ihre Potentialtheorie. IV, 175 Seiten. 1966. DM 14,– / $ 3.50

Vol. 23: P. L. Ivănescu and S. Rudeanu, Pseudo-Boolean Methods for Bivalent Programming. 120 pages. 1966. DM 10,– / $ 2.50

Vol. 24: J. Lambek, Completions of Categories. IV, 69 pages. 1966. DM 6,80 / $ 1.70

Vol. 25: R. Narasimhan, Introduction to the Theory of Analytic Spaces. IV, 143 pages. 1966. DM 10,– / $ 2.50

Vol. 26: P.-A. Meyer, Processus de Markov. IV, 190 pages. 1967. DM 15,– / $ 3.75

Vol. 27: H. P. Künzi und S. T. Tan, Lineare Optimierung großer Systeme. VI, 121 Seiten. 1966. DM 12,– / $ 3.00

Vol. 28: P. E. Conner and E. E. Floyd, The Relation of Cobordism to K-Theories. VIII, 112 pages. 1966. DM 9,80 / $ 2.45

Vol. 29: K. Chandrasekharan, Einführung in die Analytische Zahlentheorie. VI, 199 Seiten. 1966. DM 16,80 / $ 4.20

Vol. 30: A. Frölicher and W. Bucher, Calculus in Vector Spaces without Norm. X, 146 pages. 1966. DM 12,– / $ 3.00

Vol. 31: Symposium on Probability Methods in Analysis. Chairman. D. A. Kappos. IV. 329 pages. 1967. DM 20,– / $ 5.00

Vol. 32: M. André, Méthode Simpliciale en Algèbre Homologique et Algèbre Commutative. IV, 122 pages. 1967. DM 12,– / $ 3.00

Vol. 33: G. I. Targonski, Seminar on Functional Operators and Equations. IV, 110 pages. 1967. DM 10,– / $ 2.50

Vol. 34: G. E. Bredon, Equivariant Cohomology Theories. VI 64 pages. 1967. DM 6,80 / $ 1.70

Vol. 35: N. P. Bhatia and G. P. Szegö, Dynamical Systems. Stability Theory and Applications. VI, 416 pages. 1967. DM 24,– / $ 6.00

Vol. 36: A. Borel, Topics in the Homology Theory of Fibre Bundles. VI, 95 pages. 1967. DM 9,– / $ 2.25

Vol. 37: R. B. Jensen, Modelle der Mengenlehre. X, 176 Seiten. 1967. DM 14,– / $ 3.50

Vol. 38: R. Berger, R. Kiehl, E. Kunz und H.-J. Nastold, Differentialrechnung in der analytischen Geometrie IV, 134 Seiten. 1967. DM 12,– / $ 3.00

Vol. 39: Séminaire de Probabilités I. II. 189 pages. 1967. DM 14,– / $ 3.50

Bitte wenden / Continued

MIX
Papier aus verantwortungsvollen Quellen
Paper from responsible sources
FSC® C105338
FSC
www.fsc.org

If you have any concerns about our products,
you can contact us on
ProductSafety@springernature.com

In case Publisher is established outside the EU,
the EU authorized representative is:
Springer Nature Customer Service Center GmbH
Europaplatz 3, 69115 Heidelberg, Germany

Printed by Libri Plureos GmbH
in Hamburg, Germany